THE WOUNDED EARTH

THE WOUNDED EARTH

AN ENVIRONMENTAL SURVEY

by CARL MARZANI

YOUNG SCOTT BOOKS

In Memoriam

Edith Eisner Marzani

Text Copyright © 1972 by Carl Marzani
All Rights Reserved
Young Scott Books
Addison-Wesley Publishing Company, Inc.
Reading, Massachusetts 01867
Library of Congress Card Catalog No. 77–173784
ISBN: 0–201–09412–6
Printed in the United States of America
Designed by Emma Landau
Third Printing
BP/BP   8/73   09412

◐ CONTENTS

ACKNOWLEDGMENTS

The author wishes to express his thanks to Mrs. Millicent Selsam and Professor Paul Shepard, who graciously reviewed the manuscript and gave many valuable suggestions.

THE WOUNDED EARTH

⬤ INTRODUCTION ⬤

As the American astronauts hurtled through the void to the moon, they were excited and exhilarated to see Earth floating in space. Over their radio came the cry: "The earth is one." Thus earthlings, for the first time in world history, confirmed with their own eyes the lesson that ancient sages sought to impart: indeed, the earth is one, a single inextricable whole of rocks and minerals and air and water and plants and animals—including man himself.

It took millions and millions of years for the earth to evolve to the point where the balance in the mineral kingdom of rocks, water, and air could support the vegetable kingdom, more millions of years for the animal kingdom to evolve. In this sequence, man is a very late arrival—how late is hard to conceive without some aid to the imagination.

If you take the height of the Empire State Building (1,251 feet) as representing the age of the earth, you have a scale where a foot represents four million years. The TV tower (200 feet) represents the period for which we have a continuous geological record, including fossils of plants and animals. Now stand a small egg (boil it first) on top of the TV tower. It will be invisible from the street, but

its height of less than two inches represents the entire existence of *Homo sapiens*. The thickness of the eggshell (a
little more than the paper of this page) represents all of
recorded human history. The modern era of science, say
from Galileo's time, is represented by a fleck of dust on the
egg.

This very late arrival on earth, man, began early in his
career to interfere with the world around him but, for a
long time, his impact on the environment was limited by
his weakness, both in numbers (population) and in skills
(technology). Even in prehistoric times, however, it is
probable that man managed to eliminate many animal
species. For example, some paleontologists believe the
mammoth was eliminated by man. Within historic times
(the thickness of the eggshell on the TV tower) man's impact on the environment is more clearly established. For
example, he created deserts in many areas by denuding
forests. But it is in the last few centuries (in that speck of
dust on the eggshell) that scientific man has developed the
technology that increasingly allows man to interfere on a
massive scale with the processes, relations, and balances that
evolved in nature over millions of years.

This interference is necessary and proper. There is
nothing sacred about nature, and man cannot leave nature
alone. From the moment he domesticated animals and
planted crops, he was interfering with nature on a large
scale, in some cases, disastrously. Yet, without agriculture,
men could not have moved from barbarism to civilization,
could not have developed himself. He should not, and cannot, go back.

Ancient agriculture up to quite modern times was based
on slavery or serfdom. Only a thin ruling layer at the top

could be really civilized. Industrialization removed the necessity of slavery, and contemporary technology makes it possible, in principle, for all men to be civilized.

The problem, therefore, is not whether or not to interfere with nature, but to interfere in such a way that the consequences are not disastrous to man. The reason this problem has become acute recently is because the changes made by man are so vast and so rapid that natural processes cannot cope with them. Fundamentally, it is the rapidity of change that is causing trouble. Change is inherent in nature as in society, so that all relations, balances, and equilibriums are constantly readjusting and being renewed. But generally the change is so slow that animal and plant life adapt. Even such drastic changes as the great ice ages, when sheets of ice a mile thick covered much of Europe, North America, and Northern Asia, came so slowly that man had time to retreat before the ice and adapt to a colder climate. But now the changes introduced by man are so rapid that the results can be catastrophic.

The solution is not to refrain from change, but to know and understand the consequences of our actions. The harm done comes from ignorance rather than evil intent. In fact, damage often occurs in the pursuit of beneficial goals, as in the antimalaria campaign on the island of Borneo.

Malaria is a dreadful disease transmitted by certain female mosquitoes. Kill the mosquitoes, and you reduce the disease until it is finally eliminated. A powerful insect-killing chemical, DDT, was sprayed on the huts of peasants in Borneo, and as expected, killed mosquitoes, thus reducing malaria. But the spraying triggered a series of events that nearly killed a lot of peasants through an entirely different disease.

Here is what happened. The amount of DDT used in the spray is very, very small—just enough to kill the tiny mosquitoes. Roaches in the huts also ingested the spray, but being so much bigger, were not affected by the DDT. The roaches were eaten by lizards which suddenly became sluggish, slow to react and run, so that cats found it easier to catch and eat them. Suddenly, the cats died. The reason, discovered later, was that DDT accumulates in the body tissues, first in the roaches, more in the lizards (enough to affect their nervous system), and most of all in the cats. The cats died from DDT accumulated in the lizards that had eaten the roaches that had eaten DDT that had been sprayed in the huts that the peasants had built.

The destruction of the cats was a near-disaster. Rats from the forests invaded the clearings and the huts. Without cats to kill them, they began to multiply; and since rats carry the terrible plague, the danger of an epidemic became acute. A plague epidemic would kill many more people than malaria could.

With great urgency, cats were parachuted (literally) into the villages to catch or drive away the rats. When this was done, everyone breathed a sigh of relief. Suddenly, another thing happened.

Mysteriously, the roofs of the huts began to cave in, here, there, everywhere. It turned out that the lizards had been eating not only roaches, but also the caterpillars which fed on the thatching and roof beams. With the lizards dead, the caterpillars multiplied and ate into the roof beams to such an extent that the beams collapsed and the roofs with them. In due course, the roofs were repaired, and man's miscalculation could be remedied. Often, it cannot be.

It takes 500 years to build up an inch of topsoil. If

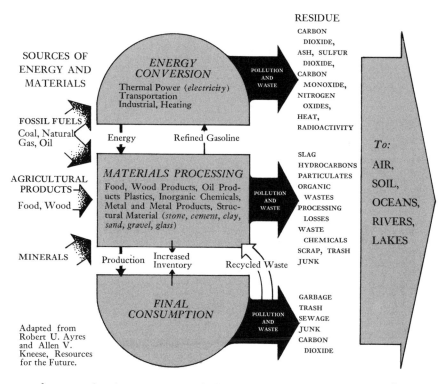

RESIDUE

SOURCES OF ENERGY AND MATERIALS

ENERGY CONVERSION
Thermal Power (*electricity*)
Transportation
Industrial, Heating

POLLUTION AND WASTE

CARBON DIOXIDE, ASH, SULFUR DIOXIDE, CARBON MONOXIDE, NITROGEN OXIDES, HEAT, RADIOACTIVITY

To:

FOSSIL FUELS
Coal, Natural Gas, Oil

Energy   Refined Gasoline

MATERIALS PROCESSING
Food, Wood Products, Oil Products Plastics, Inorganic Chemicals, Metal and Metal Products, Structural Material (*stone, cement, clay, sand, gravel, glass*)

POLLUTION AND WASTE

SLAG HYDROCARBONS PARTICULATES ORGANIC WASTES PROCESSING LOSSES WASTE CHEMICALS SCRAP, TRASH JUNK

AIR, SOIL, OCEANS, RIVERS, LAKES

AGRICULTURAL PRODUCTS
Food, Wood

MINERALS

Production   Increased Inventory   Recycled Waste

FINAL CONSUMPTION

POLLUTION AND WASTE

GARBAGE TRASH SEWAGE JUNK CARBON DIOXIDE

Adapted from Robert U. Ayres and Allen V. Kneese, Resources for the Future.

farmers, lumbermen, dam builders, and others, because of ignorance or greed, create conditions where that topsoil is washed off or blown away, it would take many generations to repair that damage.

Damage to man's environment takes many forms, but it is an overall, earthwide problem. It can be examined piece by piece for clarity of understanding and ease of research, but one must never forget that each aspect is related to all the others. Water, air, and thermal pollution, squandering of resources, the degradation of the environment, whether on farm land or in urban ghetto, the problems of garbage, noise, radioactivity: all are inextricably intertwined in one fundamental problem—man's relationship to his environment, to the planet Earth in all its aspects.

When the spaceship on the way to the moon blew out an oxygen tank and the astronauts had to improvise their self-rescue, using the lunar module as a "lifeboat," everyone was conscious of the limited amount of oxygen, water, fuel, and electric power which the module carried. The world watched the tense drama as the astronauts carefully rationed their resources. Meticulous planning and thrift became essential to survival.

We live on Spaceship Earth, and its resources are limited. If we mindlessly end up with too little oxygen, as we may, or too much carbon dioxide, as we may, or not enough clean water—and we may—there will be no "lifeboat" to take us to a safe haven. It behooves us to stop acting as high-living, wasteful passengers on Spaceship Earth and realize we have the responsibilities of a crew. We must know the consequences of our actions on our planet; we need to think ecologically.

Ecology is the study of the interrelations of living things and their environment. Everything on earth is interconnected and interdependent: all the earth's rocks, waters, air, plants, and animals form one unbroken whole. This intricate web is called an ecological system, or ecosystem for short. For convenience, ecologists divide the earth according to climate, which controls vegetation, which in turn controls animal life. These areas are called biomes, and they embrace the broadest ecological systems, such as the ecological system of a forest or of a desert. Within each of these systems we can isolate, for convenience, smaller systems. Thus we speak of the ecology of a pond, or of a river, of the ecosystems of plant life or fish life in a pond, and even of the ecosystem of a particular species. We can even speak of the ecosystem of a single plant or a single animal, and we mean by that its relationship to the whole of its environment, including other plants and animals. In the case of animals, this includes their interior organisms, such as intestinal bacteria.

Ecologists are the scientists who study ecosystems, and the term is applied to those who study specific ecosystems which are fairly clearly defined: the ecosystem of a lake, of a plant or animal species. When the subject matter is very broad, embracing various large ecosystems as well as political and economic actions in regard to ecosystems, the person is called an environmentalist. Quite often, environmentalists tend to be social scientists, in contrast to ecologists, who are physical scientists. This book, for example, has been written by a professional economist who has some claims as an environmentalist and hardly any as an ecologist.

In common usage, ecology tends to be used in its broadest meaning, drawing upon all the physical and social sciences. It is thus an exciting, interdisciplinary science, and more and more people turn to it as they become concerned with the quality of the world around them. The depth of this concern was shown by the great success of the first Earth Day, April 22, 1970. This huge national demonstration and teach-in was sparked in the main by high school and college students applying the knowledge gathered by their teachers and elders over many decades of research and study.

The message of Earth Day is that there is a crisis of survival for mankind. The phrase "a crisis of survival" is a very strong expression and, if true, represents a very frightening fact. Many political and industrial leaders however, think it is symbolic rather than factual, and that therefore the phrase is too alarming. On the other hand, many well-qualified scientists think it is literally true. Why the difference?

The answer hinges largely on the meaning of the word "crisis" which is usually associated with a short period of time. Everyone would agree that a thermonuclear war, lasting perhaps half an hour, would constitute a crisis of survival. But in relation to nature, we have to think of much longer spans of time as being critical—perhaps as much as a hundred years. This is what the scientists have in mind. They say that if nothing is done, or not enough is done, to stop the current rate of harm to the environment, then there won't be enough coal for the population of the United States within five generations, not enough topsoil in four, not enough clean water in three, and no way to breathe city air in two. For mankind as a whole, one third to one half of the next generation might die of starvation or malnutrition—unless something is done.

The phrase "crisis of survival" is not an abstraction. One of the great world scientists, Dr. Albert Szent-Györgyi, Nobel Prize winner for his discovery of vitamin c, recently put the chances of man's survival at no better than 50 per cent and steadily dropping. "The question is," he said, "which course will man take? Toward a bright future or toward exterminating himself? At present, we are on the road to extermination." (*New York Times*, February 20, 1970.) This is the actual, factual, continuing situation: it should be, it *is*, frightening.

Part I

THE ENVIRONMENT

This section has been enormously compressed to give a brief orientation with which to understand the magnitude of some of the problems of pollution. Two factors must be kept uppermost as one reads the following three chapters: the limitless complexity and interdependence of all ecosystems, from the tiniest to the largest, and the awesome time scale over which they have evolved. This is why it behooves man to treat the environment with caution. He rarely knows the consequence of his actions and he can, in one act, destroy a balance that took a hundred million years to evolve.

◑ I ◐

HOW THE ECOSYSTEM EARTH BECAME WHAT IT IS

Wherein we speak of the earth's beginning and how there were no rocks, no water, no life, no air, and how each came to be. The tricky balances that had to develop in nature to give rise to life and its supporting biosphere. We refer to the "infinite" sky and the "bottomless" oceans and find there is no such thing.

Ecosystems may be thought of as balance points of impinging natural (and human) forces always in a process of change, disruption, and renewal. These balances may be so precarious as to last a fraction of a microsecond (a millionth of a second) or so stable as to last a thousand years—a microsecond in stellar time. In creating the ecosystem earth, balances evolved over millions and billions of years.

Some five billion years ago the earth was born as a mass of hot gases spinning in space, spinning and cooling. As it cooled the gases condensed into a fluid state, until the earth was one huge molten mass of minerals with a thick atmosphere of gases and water vapor.

The fluid earth was incredibly hot. In the viscous fluid, currents were set up, with the heavier elements, such as iron, sinking to the center, while the lighter materials, such as granite and basalt, were pushed to the surface.

Over millions of years the earth cooled until a thin crust began to form. This crust cut off some of the interior heat, so the process of cooling became progressively faster as the crust became thicker, until finally the thick crust blocked the interior heat completely and the temperature of

the crust was determined by solar radiation. Above the cooling crust, water vapor condensed and fell as rain. At first the
crust was so hot that the rain quickly evaporated, further
cooling the crust. Little by little some water stayed, running
down the higher parts of the earth to the ocean basins.
It took a long, long time to fill up the ocean basins.

As the rains fell and eroded the rocky earth, they carried
minerals into the sea. One of these minerals is salt, which
dissolves in water, so century after century the oceans became more and more salty until today seawater is 3 per
cent salt. The amount of salt is enormous: more than 40
million billion tons (40,000,000,000,000,000)! If distributed
on land, it would come to a layer 450 feet high! It is estimated that it took about one and a half billion years to
erode this much salt into the oceans.

We conclude that some two billion years ago the earth
was stabilized into a rough division of rocks, oceans, and
air, but it was an earth that we wouldn't recognize. The
continents and the oceans that we know, our present mountain ranges and rivers either didn't exist or were in different
places. Above all there was no green mantle over the land,
no plants or animals or fishes—in fact, there was even no
air as we know it today.

Today's air consists of 78 per cent nitrogen, 21 per cent
oxygen and the remaining 1 per cent is a smattering of
other gases including .00055 per cent hydrogen and
helium. At birth, however, there was no free oxygen in the
atmosphere, a tiny percentage of nitrogen, and an enormous
amount of hydrogen and helium—about 99 per cent. What
happened to reverse these percentages?

One of the fascinating things we know about the universe as a result of long and intensive studies is that the

chemical composition of matter is uniform throughout space—it is the same everywhere in the universe. About 55 per cent of cosmic matter is hydrogen, 44 per cent is helium (or 99 per cent together), and the remaining 1 per cent covers all the rest of the known elements—iron, nitrogen, uranium, etc.—of which we know a hundred-odd. Scientists assume that this distribution of chemicals has remained the same throughout the eons of time. Therefore, in all probability, the earth at birth had the same composition as cosmic matter, or 99 per cent hydrogen and helium.

The dispersal of hydrogen and helium is a grandiose example of natural forces working out new balances in nature. They are very light gases, and the earth's gravity (as it cooled, shrank, and lost speed) could not hold them. They escaped into space. As they escaped, the nitrogen— held by gravity and remaining behind—increased in importance. Originally, its *percentage* was tiny in comparison to hydrogen and helium, but its *volume* was substantial. The volume of nitrogen remained about the same; the other two gases practically disappeared. As a result, nitrogen's percentage of all gases increased to its present 78 per cent. Oxygen, once nonexistent in a free state, was being created by plant life, and increased in volume (and percentage) until it became stabilized at 21 per cent. Together, therefore, oxygen and nitrogen now represent 99 per cent of the atmosphere, supplanting hydrogen and helium.

A simple way to visualize what happened is to think of a football game with 100 players (most of them on the bench, of course) and 10,000 spectators. The players are 1 per cent of all people present in the stadium. At the end of the game all the spectators, save one, leave while the 100 players hang around for a few minutes. The players are

now (roughly) 99 per cent of those present, and the
spectator is 1 per cent although the number of players
hasn't changed at all. Just as the number of spectators was
very large compared to that of the players, the amount of
hydrogen in the original atmosphere was enormous, so
enormous that it took millions of years to escape into space.
Moreover, although very light, hydrogen does have weight,
and its enormous quantity pressed down so that during
those millions of years the gas was under great pressure.

What followed is again a grandiose example of repercus-
sions and interdependence of natural balances. For in those
millions of years, the presence of hydrogen under pressure
served as the foundation of all life on earth.

That plant and animal life developed from inert matter
is accepted by nearly all leading scientists today. We still
find transitional forms. For example, the tobacco virus has
a crystalline form typical of inert matter, but under certain
conditions it shows aspects of life: it takes food and reproduces. However, the precise steps by which matter became
life are still in the process of discovery and verification in
laboratories. The leading theories rely on the enormous
quantities of hydrogen then existing in the atmosphere as
the foundation stone of the process which led to life.

The reason is this: hydrogen under pressure, and its
simple compounds (methane, ammonia, water), when exposed to high energy sources, such as lightning or the
ultraviolet rays of the sun, give rise to complex molecules
known as *amino acids*, which, in turn, are the key components of extremely complex molecules known as *proteins*, which are the core of living organisms.

By a series of chemical reactions (most of which have
been reproduced in laboratories), microorganisms which

did not depend on sunlight or organic stuff for food appeared in the oceans. These are "mineral-eating" organisms, such as still exist in the "sulphur and iron bacteria," which obtain their energy from the oxidation of sulphur and iron compounds. It is believed, incidentally, that the thick deposits of bog iron ore, the main source of iron in the world, are the work of these iron bacteria over thousands of millions of years.

Over time, the earth grew cooler, the oceans larger, the heavy clouds thinned out, and sunlight fell on the waters. Some primitive organisms developed substances that could absorb sunlight. These are known generically as photosynthetic pigments, of which the best known is the green chlorophyll. Red and blue algae contain pigments known as phycobilins, and many higher plants have a third type of pigment known as beta-carotene. These pigments make possible photosynthesis, the process of combining sunlight with water and air to form the food of the plants. Light splits the water (H_2O), and the hydrogen is added to carbon dioxide in the air to form carbohydrates, while the oxygen from the water is used by the plant for its energy requirements and also is released into the air in periods of light. It is believed that all oxygen in today's atmosphere has come from plants.

While some microorganisms "fed" from air and sunlight and thus are the ancestors of plant life, other microorganisms began to feed on them (and soon on each other), becoming the ancestors of animal life.

In the oceans today are billions and billions of tiny organisms, many of them one-celled and most of them microscopic, which are the lineal descendants of those primitive organisms we've mentioned. They are extremely diverse, but they are categorized for convenience into zooplankton if animal and phytoplankton if vegetable. The phytoplankton support all life in the seas—the untold billions of fish—and any cutting down by pollution is potentially disastrous to man.

Besides the permanent planktonic life, there is the temporary addition of such organisms as the eggs of fish. This is not negligible: a single cod may spawn several million eggs annually. The expression "planktonic life" refers to the totality of tiny organisms within the photic (light-penetrated) zone of the oceans.

The basic division between animal and vegetable forms took place perhaps 500 million years ago and then slowly evolved into the present balance between the two. This balance rests on other variables, such as the composition of the air, the heat on the earth's surface, the amount of water available, and so on, all of which are in balance with each other and have achieved that balance over billions of years.

Air and water are crucial to life and form a thin envelope around the earth which is known as the *biosphere* —the most complex of ecosystems. This envelope is very thin, for while there are traces of air as high up as 70 miles, more than 80 per cent of all the air is in the first 10 or 11 miles. Similarly, though the greatest depth of water is seven miles (in the huge Mindanao trench at the bottom of the Pacific Ocean near the Philippine Islands), more than 90 per cent of all the water of the earth is within two miles of the surface.

Practically speaking, therefore, the biosphere is about 13 miles thick or 1/300th of the radius of the earth, about the same ratio as the thickness of its cover is to a basketball.

Whatever causes damage to water and air damages all

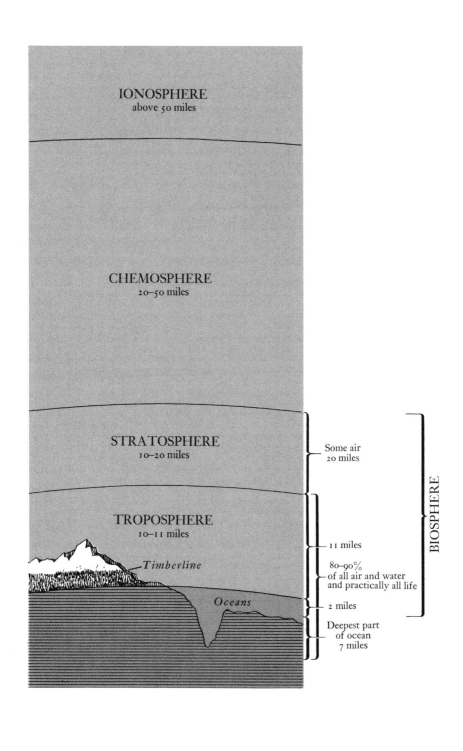

IONOSPHERE
above 50 miles

CHEMOSPHERE
20–50 miles

STRATOSPHERE
10–20 miles

TROPOSPHERE
10–11 miles

Timberline

Oceans

Some air
20 miles

11 miles

80–90%
of all air and water
and practically all life

2 miles

Deepest part
of ocean
7 miles

BIOSPHERE

life, including man. Whatever causes drastic changes in
the composition of water and air, or in their relationships,
damages all life, including man. In the history of the earth
there have been many changes, some very great. For example, there have been recurring ice ages, due perhaps to
changes in the temperature of the air. Many species of
plants and animals must have perished, and man was forced
to move away from native habitats. Volcanoes exploded,
hurricanes raged, tidal waves roared—all with great damage
to life in their given areas. But, in general, these natural
catastrophes were widely separated in time and space and
were not cumulative. Man's interference with nature has
been very different. Increasingly, his activities have spread
over the entire earth, and have been going on steadily,
year after year, decade after decade, and above all have
tended to be cumulative in many areas. Man's impact today
is often planetary, and one of our major problems is our
ignorance of the consequences of our actions.

For example, take temperature as a factor in the ecosystem earth. The temperature of the earth's surface is
fairly stable. If it got much cooler, we would have an ice
age with much of the western world under several miles of
ice; if it got much warmer, the polar ice caps would melt,
raising the level of the oceans about 200 feet, wiping out
the major coastal cities.

With this in mind, consider the increase of carbon
dioxide in the air as a result of combustion of coal, oil,
wood, etc. How big the increase is we don't know; some
people guess a 25 per cent increase over the last 100 years.
Carbon dioxide acts to trap heat near the earth's surface
as a result of the so-called "greenhouse effect." The glass
of a greenhouse is largely transparent to the short wave-

lengths (visible light) from the sun. The contents of the greenhouse are warmed by this radiation, and in turn radiate heat in longer wavelengths (infrared light). This radiation is partly absorbed by the glass and partly re-radiated (about half) back into the greenhouse. So a greenhouse is warmer in the daytime than the surrounding air.

Without the "greenhouse effect," the temperature of the earth could be as low as 7 degrees Fahrenheit and everything would be frozen. With the increase in carbon dioxide, could the earth be overheating?

Or, alternatively, could it be overcooling? The earth is heated by solar energy, and whatever affects the amount of solar energy reaching the planet directly affects its heat balance. A great deal of solar energy is reflected back into space, and the percentage of this reflection is called the *albedo* (from the Latin *alba*, white). Clouds, dust particles, ice caps, deserts—all help reflection, which may run from 30 to 60 per cent on any given day. There is also absorption by clouds and dust, which may run from 5 to 20 per cent.

Particles of matter—dust, smoke, etc—act both as a shade to the earth and as "condensation nuclei" around which water vapor gathers to form clouds. A similar effect is produced by the contrails of jet airplanes and is one of the objections ecologists have to the supersonic plane. Increasing cloud cover sharply increases the albedo, so that on a cloudy day less than 45 per cent of solar energy reaches the surface of the earth—under exceptional circumstances almost none. Cloudy days are of course offset by clear days when as much as 70 to 80 per cent of solar energy reaches the earth's surface. The present balance

gives an annual average of 57 degrees Fahrenheit. The
ever-increasing pollution is bound to affect this balance,
and either direction (whether of heating or cooling) can
be catastrophic. Will the polar caps increase and usher
in an ice age, or will they melt and drown the coastal
cities?

No one knows, for we have no direct and precise
measurements of atmospheric temperatures in the distant
past. There are some very good estimates of ocean
temperatures based on the chemistry of shells, but they
are not adequate for the issue involved. In the last 30
years accurate measurements have been made, and they
seem to indicate a slight drop in temperature, but the
period is too short to warrant firm conclusion. What we
do know, however, is that the seemingly limitless and
self-purifying atmosphere will be, in fact, subject to definite
limits of pollution if it is to remain beneficial to life on
earth.

As with air, so with water. Water, too, seems limitless,
and the quantity of water in the oceans is indeed tremendous—314 million cubic miles. If the earth had no
huge craters, such as the Pacific basin, and the oceans
were distributed over the surface, every bit of the earth
would be under 9,000 feet of water. Only the highest
mountain peaks would stick out like islands. Yet here,
too, as with the air, the enormous quantities are deceptive.

The supply of fresh water on earth is much less than
that of salt water, and much of it is not readily available:
it is locked up in the polar ice caps or in deep aquifers
—the name used for layers of water beneath the surface.
What is readily available is the water in lakes, ponds and
rivers. Some of the water from shallow aquifers is tapped

THE WATERS OF THE EARTH

(in cubic miles)

* Comprises rivers, lakes, ice, and all aquifers

** A deep aquifer is from 2,500 feet to 15,000 feet below the surface.

*** Surface waters and aquifers to be a depth of 2,500 feet.

**** Lakes, ponds, and rivers. Rivers only contain 75 cubic miles at any given time.

TOTAL SALT WATER

(oceans, seas, gulfs, etc.)

TOTAL SWEET WATER *

POLAR ICECAPS

DEEP ** AQUIFERS

***

****

| 314,000,000 | 10,000,000 | 6,000,000 | 3,000,000 | 1,000,000 | 30,000 |

through wells, and some of the underground water feeds lakes and streams, which total some 30,000 cubic miles, of which one fourth is in North America. Since a cubic mile of water is roughly a billion gallons, this is still a lot of water, though it is very clearly limited. In an advanced country such as the United States, the use of water has escalated to such an extent in the past hundred years that we are beginning to press against the limits of sweet water.

Rainfall is the only source from which underground waters, lakes, and rivers can be replenished. The total rainfall in the United States is estimated at 4,300 billion gallons a day, of which some 70 per cent returns to the atmosphere by evaporation, animal perspiration, and its plant equivalent—transpiration, whereby water vapor works out to the surface of leaves and evaporates. A tree

may transpire as much as 80 gallons of water a day.

This leaves some 30 per cent, or 1,300 billion gallons a day, but at least half of this is not accessible for various reasons: it is far from population centers, it percolates into the very deep aquifers, and so on. The dependable water supply in the United States is therefore around 650 billion gallons daily. We are now using 400 billion gallons a day, and this figure is steadily rising, year by year. It is estimated to reach 700 billion in 1985, which means that within a dozen years we will be using more water than nature replaces. Unquestionably, we can increase our water supply, but at what cost? Furthermore, we have been speaking as if all this water were clean water, which is hardly the case.

The problems of the United States, though perhaps more acute, are duplicated in all industrial nations. Even the Soviet Union, with its huge area and mighty rivers, is facing a clean water shortage. A special research body was set up in 1967, the Institute of Water Problems, and in May 1971 its chief, Professor A. N. Voznesensky, warned that planning studies were not getting sufficient support to meet the expanding water needs of the country. He estimated that water needs would more than double over the next fifteen years and they couldn't possibly be met unless drastic measures were taken. As reported in the *New York Times* (May 16, 1971), he warned that sewage treatment facilities alone couldn't do the job of protecting streams but that the solution had to be found in recycling water and burying the most harmful or concentrated wastes.

Earthlings must face the fact that both air and water are limited in amounts, as well as limited in self-regenerative powers. In many cases, we are pressing against those limits. Moreover, it took billions of years to achieve the present balances within the biosphere. They are not to be lightly disregarded: we can upset in a century the work of millennia. In some ways the biosphere is very tough; in other ways, it is very fragile. While there is no cause for panic, there is cause for concern—and circumspection.

◐ 2 ◐

THE IMPORTANCE OF BEING AN EARTHWORM

Wherein we talk of topsoil—the factory of our daily bread—of rocks and lichens and roots and erosion which created that topsoil. How the topsoil is plowed and renewed by earthworms. A small argument that the earthworm, not the dog, is man's best friend, to be loved—if not taken to bed.

Rocks, air, water, and sunlight made the miracle of microorganisms.

Rocks, air, water, sunlight, and organisms made the miracle of topsoil.

Air, water, sunlight, and topsoil made possible the tremendous spread of plants over the earth and thus the higher forms of animal life became possible.

Without soil, life would be limited to water; yet, without life, there would be no topsoil. The microorganisms born two billion years ago made topsoil possible. For sand is not topsoil, and neither is dust. Topsoil is an ecosystem —a balanced combination of rock particles and organic matter (such as decomposed leaves) plus microorganisms (such as soil bacteria). If you go to a park and dig up a spoonful of dirt, it will tend to be moist, dark, and crumbly. It looks dead, yet that spoonful contains millions and millions of microorganisms—bacteria, fungi, algae. It also contains millions of fine particles of matter—dust. We have seen how microorganisms arose; how did rocks get ground down into such fine particles?

Erosion is the name given the process by which the pressures of wind and water constantly break particles off rocks. Erosion is furthered by changes in temperature. Sunlight falls in unequal amounts at various times and on various parts of the earth, largely depending on the way the earth turns on its axis and orbits around the sun. This results in the seasonal and diurnal variations of heat and cold. Heat and cold work with water and air to break up rocks. Heat expands rocks, cold contracts them. The combination causes cracks. Rainwater runs into the cracks, the water freezes, the cracks expand, and the rocks flake and disintegrate. Air contains carbon dioxide; the falling rain dissolves CO_2 into carbonic acid, which slowly, slowly eats out some minerals in the rocks, making more crevices for more water to enter and freeze and widen cracks. Boulders lose their footing and go hurtling down slopes, grinding against other rocks. Torrents carry rocks along, abrading against each other. Winds erode surfaces. Mountain glaciers slowly move down clefts and valleys, grinding the rocky sides. During the ice ages, enormous glaciers spread and ground their way over large areas of the earth. Over millions of years, the grinding of rocks into particles went on.

While rock particles were slowly accumulating, plants —extinct forms of algae, mosses, etc.—were invading the land. The shift from water to land was a most difficult one. Once on land, plants had to develop stems. They had to develop root systems and inner channels to bring water from roots to leaves and soluble food from the leaves to the rest of the plant, including the roots. The beginning of the plant invasion of land probably took place in shallow coastal water, where the receding tide would leave plants dry for shorter or longer periods of time.

When we speak of the invasion of land by plants, we don't mean any plan on the part of plants to get onto the land. Nor do we mean adaptation for the purpose of getting on land. What happens is that organisms desperately cling to life. If receding waters leave a plant or a fish behind, it will try to stay alive as long as possible until the waters return. Most members of a species will die, but if any member is just a little different in some small way that helps it to survive until the water returns, that member will survive and propagate. Thus a fish whose gills can take just a tiny amount of oxygen from the air, rather than from the water, may survive. Over millions of years genetic changes take place, and through natural selection new species evolve.

The detritus from sea life fertilized the coastal areas, and the invading primitive plants found the soil ready for them. Some may have helped make the soil, just as the lichens do to this day. Lichens are tiny leafless plants without roots which make acids from the air. These acids dissolve some of the minerals in the rocks, making crevices where dust and dead lichens accumulate. Slowly, slowly over the years dust turns to soil, the lichens build up a seedbed for other less hardy plants, such as ferns, which take root, grow, and die, adding their organic material to the seedbed which finally grows thick enough to support shrubs and seedlings of trees. The plants grow to seed, the seeds scatter to make new plants, and the parent plant dies. The dead plant's roots and leaves provide food for tiny organisms—bacteria, molds, etc.—which decompose the dead plant into chemical forms suitable as food for new plants.

The decomposition attracts all kinds of small creatures

that feed on the decomposing material which includes bacteria. The importance of these small creatures cannot be exaggerated, not only in enriching the soil, but in keeping the bacteria in check. Soil bacteria are among the simplest of living things and extremely small. A teaspoon would hold millions. They reproduce by splitting (one individual into two) perhaps twice an hour. At the end of 24 hours, one bacteria would give rise to 17 million of them, and by the end of a week their bulk would exceed that of the earth. Unless kept in check, every living thing would be fighting for air and space.

Among these tiny scavengers, the lowly earthworm holds pride of place. No poet has sung its praises and no monuments have honored him, yet he might well qualify as man's best friend. For the earthworm is a great farmer.

The earthworm is a walking stomach and intestine combined. It ingests earth at one end of its hollow body, pushes it through and out the other end as castoff or castings. The earth includes organic materials and bacteria which are thoroughly mixed and digested by the earthworm and the castings are often left on the surface. The earthworm burrows down vertically and brings the deep earth up to the surface, thus aerating the soil and making it easier for roots to spread. The burrows also make it possible for the soil to absorb more surface water as well as to drain the soil. Earthworms nudge small seeds underground and help germination. They work continuously. Darwin estimated that in normal circumstances they bring up some 8 to 18 tons of earth per acre each year depending on their population per acre. In some farms today farmers are cultivating earthworms commercially for distribution to other farmers and to householders who would like to have a fine lawn without using dangerous chemical fertilizers. The fertility of the Nile valley, which has been farmed for 6,000 years, has been attributed by British scientists both to the flooding of the Nile, which leaves its sediment behind, and to the large numbers of earthworms which inhabit the soil.

As the soil is enriched with organic matter, earthworms and ants multiply, and other small species increase—spiders, beetles, and so on. Some eat plants; some eat the tinier organisms and each other. Larger animals, such as shrews and moles, eat the smaller ones. The remainder of their meals decompose, and, together with their excrements, fertilize the soil.

Excrement is an important source of nitrogen, which plants need and cannot get directly from the air. Another source of nitrogen is the nitrogen-fixing bacteria. Nitrogen fixation is the general name given to any process which combines the free nitrogen in the air with other elements to form compounds. For example, nitrogen and oxygen form nitric oxide (NO), nitrogen and hydrogen form ammonia (NH_3), and so on. Plants can use some of these compounds, but not others. Thus ammonia is not much good as fertilizer, but nitric oxide is. So ammonia is combined with oxygen, and the result is nitric oxide plus water. (For chemistry students, the formula is $4\ NH_3 + 5\ O_2 = 4\ NO + 6\ H_2O$.) Nitrogen-fixing bacteria are those bacteria that can take nitrogen out of the air and combine it with other elements to make compounds that the plant can use, or those bacteria which change compounds not usable by plants into compounds that are usable. There are also denitrifying bacteria which reverse the process. When plants or animals die and decay or putrefy, ammonia is produced. There are bacteria which break down the ammonia, separate out the nitrogen and return it to the air.

Nitrogen-fixing bacteria are found chiefly in the roots of the legumes (beans, clovers, locust trees, etc.), and these plants are not found everywhere. So farmers also use manure and chemical fertilizers to bring nitrogen to the soil. Until 1914 these fertilizers were dependent on nitrogen compounds (the nitrates), but during World War 1 industrial methods were developed to fix nitrogen directly from the air so that it is very plentiful today.

A final component of good topsoil is the proper mixture of minerals; potassium, iron, sodium, sulphur, and so on. Too great a quantity of minerals can harm the soil—for example when seawater overruns the land for a length of time and then recedes, plants will not grow for many years, until the salt and other minerals are leached out of the soil. Not enough minerals will hurt plants. The quantities needed are small (in corn they range from 1.5 per cent for potassium to .5 per cent for sulphur), but they are essential. Very minute quantities are called trace elements, but without them the plant cannot flourish.

The minerals plants need are usually mixed with other minerals in the form of rocks and they must be separated. Roots and soil bacteria help to dissolve them out of rock particles. Root tips release carbon dioxide which, with water, makes carbonic acid that helps dissolve minerals. As bacteria decompose dead vegetation, carbon dioxide is also released. In decomposition many other acids are released besides carbonic acid, including citric, tartaric, oxalic, and malic. All these acids help to dissolve minerals.

The biggest factor in dissolving minerals, however, is

water eroding the rocks and carrying the minerals in solution into the soil. However; too much water can be as
harmful as too little. Too much water leaches the soil, that
is, it carries the minerals away from the topsoil. In areas of
heavy rainfall the soil tends to be seriously leached; fortunately, trees do not need many minerals, compared to
grass and crops. They do need a lot of water, which is why
forests do well in areas of heavy rainfall.

This, then, is the story of how the ecosystems of various
topsoils are formed as a result of a complex of smaller
ecosystems—the bacteria and worms and insects that make
the topsoils possible. The one striking fact in this story is
that everything must be in balance: enough water but not
too much; enough minerals but not too much; enough sun
but not too much; enough bacteria but not too much. Topsoil is a marvel of fine adjustments. It is the most precious
part of the earth's surface, more precious than gold and
rubies and uranium. Six inches of topsoil will support
many crops, and in some parts of the world, such as our
Midwest and the Ukraine, the rich, loamy topsoil is six
feet deep, and even more.

Topsoil depends on all the factors we have named, plus,
as we have seen, time—500 years per inch. One inch can
be washed away *forever* in 10 minutes by the rushing
waters of a flood; it can be blown away in one hour by
the scooping winds of a storm, as happened dramatically in
the 1930's on the great plains just east of the Rocky
Mountains, the so-called "short grass" prairie.

Screened by the mountains, the prairie receives little
rainfall, so that crops cannot survive the drought. The
short grass, adapting over the centuries to little moisture,
thrives. The rainfall was just enough to sustain the grass

and to dissolve minerals in the soil to a depth of 12 inches, without being heavy enough to carry the minerals away. Moreover the grass was strong enough to resist the winds that in stormy weather raged unhindered over the plains. The short grass was also excellent for animals, and the prairie was a natural grazing area.

In the late 1920's there was a period of increased rainfall, which made it possible to raise wheat. Tempted, the farmers plowed under vast acreages of grazing land and planted the more profitable wheat. When a period of drought came, wheat could no longer grow, the land was bare and exposed. As storms came, the unprotected land was blown away so profusely that the very sky was darkened by the dust. The once-luxuriant prairie was turned into the dust bowls from which people had to flee. The human suffering involved was very great and has been movingly described by John Steinbeck in his classic, *The Grapes of Wrath*.

The disaster set the government and the farmers to work in a long battle with nature. The topsoil that had blown away couldn't be replaced, but what was left could be protected and used as a base for reclaiming the land. The chief battle was against the winds—to block them, impede them, lessen their impact. Trees were planted and carefully nurtured until they were large enough and strong enough to make a barrier against the winds; furrows were dug at right angle to the winds' direction to slow them down; a hundred and one tricks were conceived and tried. Some worked. Slowly the impact of the winds was contained, in a great saga of human toil and human ingenuity which is related in detail in Ira Wolfert's book, *An Epidemic of Genius*.

The good earth is still the basis of most of the food

desperately needed by growing populations. Its ecological balances have an importance impossible to exaggerate. The ecologists and conservationists who fight passionately for the preservation of forests to protect watersheds and prevent raging floods, or urge the planting of new forests to act as wind barriers, are not fanatics and crackpots; they are the true realists of today—for they think of tomorrow.

◗ 3 ◖

TREES, GRASS, AND ECOSYSTEMS

Wherein we speak of plants and animals and their interdependence. Of the chains of life and the balances of nature—the ecosystems, large and small. Of roots, and how they guard the soil. Of man, and how he can destroy or protect the ecosystems.

As topsoil increased, plants spread across the continents, century after century, millennium after millennium, changing, dying, adapting, flourishing, always evolving—hundreds of thousands of species. As the grasslands spread and the forests grew, animal life evolved and multiplied in its myriad species, from the seals in the frozen north to the hummingbirds in equatorial forests. For two billion years, species evolved, failed to adapt, and died. Others evolved, adapted, and flourished. Every step in survival was one of interdependence of living things with one another, with their climate and with their terrain. A stupendous number of ecosystems evolved, some comparatively simple, others of an almost inconceivable complexity.

Ecosystems evolve over long periods of time and eventually reach a state of balance which is comparatively stable for the given conditions. Thus a forest, if climate and soil conditions remain roughly the same, evolves to a point where it will continue for a long time, constantly renewing itself in the process. A long process of trial and error went into making the stable forest. In the beginning, millions of seeds of many kinds of trees were brought to the given area by winds and by birds. Seedlings which liked strong sunlight grew, while other seeds couldn't get started. The seedlings grew into trees and in their protective shade, seedlings of other types of trees which needed shade began to develop, overtopped the trees that nursed them, and crowded the early trees to death. The newcomers may themselves have created conditions suitable for other seedlings which in turn grew and overtopped them, crowding them out. At last some species that were best fitted to the soil and climate conditions triumphed over all other species, and thus the forest was stabilized. This is called a climax forest. The orderly procession from pioneer forest to climax forest is typical of most plant communities. The complete series of changes is called a *sere*.

The adaptation of trees to their environment can best be seen on a mountainside, where both soil and climate change with increasing altitude. In the dry central California valleys, for example, there are no trees. As we go up the mountainside, the rainfall increases, and low-growing brush makes its appearance. A little higher, and there is sufficient rainfall for blue oak. Higher still, the ponderosa pine survives. Higher, the sugar pine can live, and still higher the Douglas fir, then the white fir and the red fir. The firs are more shallow-rooted than the pines, so they need more rainfall. Above them all, at 8,000 feet, come the mountain hemlock and the slender lodgepole pine. These are the hardiest trees; above them the climate is too cold. The hardy grasses take over until they cannot survive and only the rocks of the bare mountain peaks are left. Even here, some lichens cling to the rocks.

Water is a most important consideration for trees, and the adaptability of plants to water is fascinating. The giant tree cactus of the desert, the saguaro, and the date palm of the desert oasis both require a very hot climate. But the saguaro requires little water; the palm tree, a great deal. The saguaro will lose by transpiration as little as one fiftieth of a quart a day or one quarter of 1 per cent of its water, while the palm will lose 500 quarts or 90 per cent of all the water it contains.

The forest is in a constant struggle with the grassland; trees need a lot of water, while grass needs much less; but grass needs many minerals in the soil, while trees need much less. Where trees get a start, they shade the grass from the sunlight. Where grass gets a start, it takes up the moisture the trees need. According to the amount of rainfall and the composition of the soil, the borders between forest and grassland are established and stabilized.

Trees and grass protect the topsoil with their root systems. The root system of grass is an incredible thing. A study made on winter rye grass shows that a plant which grew to 20 inches in four months had developed 378 miles of roots and 6,000 miles of root hairs! This means an average *daily* growth of three miles of roots and 50 miles of root hairs! This experiment was made with a single plant in a box with two cubic feet of earth. The total surface of the plant aboveground was 51 square feet. One can imagine how tightly packed and interwoven were the roots. (See John H. Storer, *The Web of Life*, p. 31.)

Such a root system enables grass to take up every bit of moisture from the soil and to survive in semiarid areas. Root systems also hold the soil together and prevent the erosion caused by the wind on unprotected soil.

Forests, too, guard the topsoil. They are particularly important in the control of erosion due to rushing rivulets and streams. Unchecked, the power of water is enormous:

every year, 800 million tons of earth in the United States is washed down to the sea—not all of it topsoil, of course, but enough so that we should be seriously concerned. It has been estimated that since 1776 the United States has lost *one third* of its original topsoil.

When a hard rain hits the soil, it splatters the finer particles and drives them into the interstices between the larger particles, making a smooth surface over which water runs off easily. Trees, with their umbrellas of spreading foliage, break the impact of the rain, hold the raindrops on their leaves, storing them for a time, and give the earth below a chance to absorb more water. The same effect is produced by ferns, mosses, underbrush of all kinds, and particularly by the humus, the thick carpet of decomposing vegetable matter on the forest floor. The spongy humus takes a long time to build up: a hundred years for two inches.

Finally, the roots hold the soil together, preventing its disintegration. The water is slowed down, some of it is absorbed, the rest filters down to the great reservoirs underground, to replenish springs and rivulets, or to be tapped by wells for human use.

In winter, the thick humus prevents the ground from freezing; in spring, the shade keeps the snow from melting too fast. The waters of the slowly melting snow are absorbed by the unfrozen ground to further replenish the underground water.

In certain terrains, forests help prevent erosion by breaking the force of strong winds which might otherwise erode the soil. In measurements of a forest near Cleveland, Ohio, it was found that trees slowed down the speed of the wind to one tenth of the wind in the open fields in summer, and one quarter in winter. Wind and sun are major factors in evaporation, and since leaves also provide shade, the rate of evaporation in a forest may be 50 per cent less than the evaporation in an open field.

In the constantly changing world of plants and animals, a good deal of stability and balance developed. There were sudden catastrophes, volcanoes erupted, forests were wiped out, rivers changed their courses, deserts increased or retreated as climate changed; but on the earth as a whole, over the centuries, billions of ecosystems great and small found their balances and stability, even if sometimes it was a precarious stability.

Into this world man emerged as only one of myriad animal species and as little a threat to his environment as any other animal. For thousands of years he developed into a more complex social being, but still with no great destructive powers. It is only comparatively recently, say in the last 10,000 years, that he began to have an ever-greater impact on his environment. It is worth stressing how recent man is. We gave an illustration before, comparing the age of the earth to the Empire State Building. Here is another comparison, using time instead of feet as a scale. Suppose that the age of the earth (five billion years) were represented by a year. The oceans would have been born four months ago, life would have appeared three and a half months ago. Man appeared on earth an hour and a half ago, Egyptian civilization 30 seconds ago, Julius Caesar 10 seconds ago, and the United States became a nation one second ago!

It's in that last second, since the Industrial Revolution, that man has become a menace to his own species. Not that man hasn't done a great deal of damage in the past. The

cutting of forests in China and India have contributed to
the erosion of their cropland. In Japan and England, there
are few forests left. Greece, Sicily, Southern Italy, and the
whole African littoral through to the Middle East have
been thoroughly denuded of trees. The more technically
advanced man became, the more dangerous he has become.
The United States, the most technically advanced nation
in the world, has become the greater despoiler of its environment. America's industrial rival, the Soviet Union,
trying to catch up to American production, is also catch-
ing up on America's pollution, despite its government's
ecological orientation.

For example, we have seen in the previous chapter how
American farmers, chiefly out of ignorance, plowed under
the wiry, drought-resistant prairie grass and ended up by
creating huge dust bowls. Undeterred by this example, the
Soviet Union in 1954 did something similar, not out of
ignorance, but under the great pressure of the need for
more grain for its people. They consciously took a gamble
on the weather and decided to plow 90 million acres of
virgin steppes, their name for short-grass prairies. Ninety
million acres is a lot of land, equivalent to all the New
England states plus New York, New Jersey, and Pennsylvania, a land moreover without roads or established facilities
—water supplies, sewers, houses, and so on.

An army of young people, who lived for months in
tents, was mobilized, as well as thousands of tractors,
trucks, supplies of food, oil, and seed, and the campaign
was on. Some of the seeding was done by airplane. The
first crop came in and seemed an unqualified success.
The second crop was not as good. Then trouble began. The
fertility of the soil dropped sharply, there were years of

drought, and erosion began to set in. Today the Soviet Union, with scarce resources, is desperately trying to remedy its mistake. This agricultural failure was part of the reason Mr. Khrushchev lost his job.

Here is a second example of environmental damage irrespective of ideology. The Great Lakes of the United States and Canada comprise five lakes—Superior, Michigan, Huron, Erie, Ontario. Together, they form the largest body of fresh water in the world, with 6,000 cubic miles or 20 per cent of all the fresh water of the earth. Lake Baikal in Siberia is the second largest such body, with 5,000 cubic miles, or 18 per cent of the earth's supply. Both bodies of water are so enormous that they seem limitless to the ordinary man, and ordinary men, without much thought, built factories and cities around the Great Lakes, dumping their industrial wastes in them, decade after decade. The Detroit area dumps one and a half billion gallons of waste a day into Lake Erie; the Chicago industrial complex discharges several times that amount daily into Lake Michigan. We don't know exactly how much, but the Chicago complex includes ten steel mills, five oil refineries, and dozens of other plants ranging from paper mills to soap factories. Six major plants alone discharge a billion gallons of waste a day into the lake. The southern tip of Lake Michigan has become a cesspool. Lake Erie is even worse: it is a "dying" lake, for it has become a huge septic tank. Two decades ago there was commercial fishing on Lake Erie, a multimillion dollar industry. As late as 1955, there was a catch of 75,000 tons. Today there is no fishing industry because there are no fish left in Lake Erie, except for one lone species of scavenger carp which has adapted to the poisoned waters. A major effort to save the Great Lakes is now underway and will be described in Chapter 4.

Lake Baikal, too, seems limitless to ordinary man, and ordinary men have proceeded to build lumber mills and paper mills on its shores, utilizing the lumber of the forests around the lake. The damage to Baikal will be twofold, from the chemical contamination (particularly deplorable since there are species of plants and animals in the lake which are unique in the world), and from the deterioration of the Baikal basin by destruction of trees and subsequent erosion. Aroused Soviet citizens organized to save Lake Baikal and it looks as if they are succeeding. Ironically, the most telling arguments the conservationists mustered were examples drawn from damage to lakes in the U.S.

These examples deal with large ecological systems and can be multiplied by the dozen. Vastly more numerous are the small ecological systems, which are destroyed out of ignorance and greed. One simple example will suffice: the destruction of pine trees in southern Oregon by a sudden increase in the population of porcupines. The porcupine loves to chew on the bark of young pines, cutting a ring around the tree. Water cannot rise from the roots; the tree dies. Normally the porcupine is kept in check by the fisher, a cousin of the mink, who loves to eat porcupines and knows how to stalk them and kill them without getting punctured. Unfortunately, the fisher has lovely silky fur which is quite valuable, and over the years more and more fishers were trapped. The fewer the fishers, the more porcupines. The more porcupines, the fewer the trees.

Who could have dreamed that a taste for fine furs in London, Paris, or New York would have caused the decimation of the pine forests of Oregon?

When man disturbs ecological systems, new relationships evolve and new equilibriums are reached. When man cleared forests for crops, or plowed under the prairies, ecological systems developed on the croplands and pastures establishing self-renewing cycles that are fairly stable over considerable periods of time. If the self-renewing cycles are disturbed or destroyed, new cycles usually develop over a long period of time until the ecosystem again achieves stability. What is happening today however, is that these ecosystems are being changed so fast and so drastically that there is no time for new cycles to be established. Because of ignorance, damage is being done even when the intention is beneficial, as we have seen in the case of DDT and malaria control in Borneo.

Here is another example. For many years dairy farming in New York State has been based on the Holstein cow, which was suited to grazing on the irregular terrain of local pastures. The cows gave milk to the farmer and excrement to the ground, maintaining the soil's fertility so grass could grow for more grazing. It was an ecologically balanced system.

To increase the yield of milk, a new strain of Holstein was bred which did in fact increase the milk yield, but at the same time increased the size and weight of the cows. The larger, heavier cows were not as good at negotiating hilly pastures, but the higher milk yield paid farmers to buy feed for the cows and confine them to a feedlot. The cow was now a milk-producing machine, with feed input and milk output. The waste, however, instead of being distributed over large pasture areas and slowly assimilated by the soil, was now concentrated on the feed lots, and its constituent materials, such as nitrogen, were not assimilated, but were leached into the surface waters to become a pollutant. The ecological balance previously established was broken anew, and this time permanently.

At this point the reader may feel that I've been stacking the cards against all human activities, since everything that man does is bound to affect his environment. This is true, but there is no reason why the beneficial effects should not outweigh the harmful effects. One needs only to think of the vaccines which have wiped out or drastically limited human diseases, particularly infants' diseases. These vaccines are all to the good, even if they have contributed to a population explosion. The remedy obviously is not to bring the diseases back, but to keep the population stable by family planning and fewer children.

Science can contribute to the well-being of mankind without harming the environment, sometimes, in fact, improving it. This is illustrated by the dramatic story of a major agricultural discovery—the need of plant and animals for traces of certain minerals in the soil. The work was done by devoted and tenacious Australian scientists.

There is in Southern Australia a huge tract called the Ninety-Mile Desert, which was practically barren: no grass or forage crops would grow on it, nothing would grow except mesquite and scrub. Yet this "desert" was not really a desert; the rainfall was good—as much as 20 inches a year—there was plenty of water underground, and the soil seemed fine, meeting all the fertility tests then known. Pioneer sheepgrowers tried to grow grass there for their sheep, but failed. They finally gave up and moved about fifty miles away toward the coast where there was plenty of native grass. There they found another problem: the sheep they were raising ate the grass, but instead of getting bigger they became listless, stopped eating, and many of them died. They seemed anemic. Those that survived were not much good to the shepherds, for instead of producing curly, crimped wool, they produced a straight, stiff hair that wasn't marketable.

The anemia of the sheep was called "the coast disease," and in 1936 government scientists tried to discover its cause. They found that there was no cobalt in the environment. A healthy sheep needs 1/300,000th of an ounce of cobalt a day. This is a very tiny amount indeed, yet without it, sheep die. As soon as a tiny amount of cobalt was added to the salt licks where the sheep got their salt, the coast disease vanished. The sheep no longer died, but their wool was still stiff.

The scientists persevered. Six years later they discovered that another mineral was missing: copper. When a minute amount of copper was added to their diet, the sheep grew normal, curly wool. Then the scientists began wondering if perhaps the "desert" soil of the Ninety-Mile Desert needed some minerals for crops. They tested it and found no trace of cobalt, copper, or zinc. So they began a long series of experiments, adding a pinch of this and a pinch of that, like chefs of the soil, until crops began to grow. They found that oats need zinc but not copper, whereas alfalfa needs copper and no zinc. Clover, on the other hand, needs a trace of both. And so on.

The scientists concerned not only have made it possible to reclaim the Ninety-Mile Desert—a welcome addition to the patrimony of arable land in the world—but have made a great contribution to agriculture everywhere. The importance today of increasing agricultural production (without harming the environment) cannot possibly be overestimated. As we shall see, it is one of the crucial elements of world politics and world peace.

Man cannot refrain from acting on nature and there is nothing sacred about any given species, such as the house-fly or the mosquito, and nothing sacred about any given ecosystem, such as a desert. Many ecosystems can be modified without harm, for while they are fragile in some aspects, like nature itself, they are remarkably tough and persistent in others. The problem for man is to act with caution, forethought, and as much knowledge as he possibly can get. As in so many areas of human endeavor, a consciousness of our fallibility is salutary, offsetting the arrogance of power. Circumspection is the hallmark of the ecologist: when in doubt, don't do it.

Part II

MAN THE POLLUTER

Pollution disturbs, disrupts, and destroys ecosystems. At present levels and rates it can, and will, render the planet uninhabitable. But pollution is integral to industry as it functions today, and the population levels, existing and expected, cannot survive without industry. The problem, therefore, is to have industry without pollution. The next nine chapters will describe types of pollution.

◑ 4 ◑

DON'T SPIT IN THE WELL

Wherein we speak of waste and how it goes into someone's drinking water. Of sewage-disposal plants which are themselves polluters. Of the destruction of oxygen in streams and lakes and the reckless use of water in the U.S. The success of tertiary plants. A hopeful breakthrough in sewage disposal of decisive importance.

There is an old Jewish proverb that says: "Do not spit in the well, you may have to drink from it." Today in the United States most drinking water comes from lakes and streams (backed in reservoirs) and they are unfortunately much spat upon—by steel mills and paper mills and oil refineries, as well as tens of millions of toilets. Even farms contribute their share, for almost all farmers use fertilizers and pesticides, whose chemicals are dissolved in the waters that run off the land into streams or filter down to the underground waters.

This continuing pollution is very complex, with hundreds of different substances: bacteria and viruses, pesticides, weed-killers, metals of all kinds, phosphorus from fertilizers and detergents, acids from mills, factory wastes and the drainage of mines, all kinds of organic and inorganic chemicals, some of which are so new that we still don't know how they affect human health.

When not overloaded, lakes and rivers purify themselves. Sunlight kills germs and bleaches some pollutants. Other particles settle to the bottom as sediment. A good deal of waste is consumed and transformed by bacteria which use up oxygen in the process. The oxygen needed by bacteria

49

and by fish is replenished by natural aeration from the atmosphere and from aquatic plants.

When there are too many pollutants, troubles begin and multiply. One of the most serious sources of trouble is the depletion of oxygen. At the present rate of depletion, scientists estimate that by 1980 all the oxygen in all 22 river basins of the United States will be consumed during the dry season.*

Oxygen depletion in water is caused by many things: sewage, fertilizers, heat from power plants and certain manufacturing processes, and strangely enough, by automobiles. The impact of heat on water is called thermal pollution, which will be treated in a later chapter. Here we will discuss only its role in oxygen depletion.

Water contains oxygen in two forms: the oxygen bound with hydrogen in the water molecule (H_2O) and free oxygen, which is obtained in two ways: one, as oxygen released by photosynthesis in aquatic plants, and two, as oxygen from the air. As a stream tumbles on its way, or as the surface of a lake or broad river is whipped about by winds and storms, air is mixed up with the water. The colder the water, the more oxygen it can absorb; the warmer the water the less it can hold and the more its free oxygen is driven out. This is one reason why cold water tastes fresh and is bracing while boiled water tastes flat. Power plants and certain factories use great quantities of water for cooling. They get the water from a lake or river, cool the machinery, storage tanks or whatever, and return the water, now very hot, to the lake or river whence it came. The amount of heat released is so great that the temperature of the water surrounding the plant outlets

* *Spillhaus Report*, National Research Council, Publication 1400, 1966.

may be raised by as much as 20 degrees: under exceptional circumstances, even more. This, by the way, is enough to kill most fish.

Another major source of oxygen depletion in waters is the chemical components of fertilizers leached out of the soil. Fertilizers are rich in phosphates and nitrogen which provide foods for plants, thus greatly increasing crop yields. As a result there has been a massive increase in the use of fertilizers, from 9 million tons in 1940 to nearly 40 million tons in 1970. But the more fertilizer is used, the less effectively does it combine with natural organic matter in the soil, because in the process the soil is physically altered, especially in its porosity to oxygen. Consequently, the crop uses up a steadily smaller portion of the added fertilizer, and the residue leaches out of the soil into rivers and lakes. Thus a vast amount of phosphates and nitrogen—plant nutrients—are added to the water. The process of adding nutrients is known technically as eutrophication.

Eutrophication is a splendid example of the problem of balances in nature. A certain amount is essential to purify streams and lakes; an excessive amount destroys the purification process and pollutes the waters.

Under normal conditions, organic matter in water—twigs, leaves, dead insects, etc.—is decomposed by bacteria which use up oxygen and release carbon dioxide. Aquatic plant life uses the carbon dioxide and releases oxygen. Fish eat plants, other fish (or man) eat the fish, their remains and their waste is decomposed by bacteria, and the cycle is renewed. The ecosystem is stable.

Suddenly eutrophication is sharply increased. The nitrogen and the phosphates vastly increase the growth of algae and other plants to huge proportions. Fish cannot eat the

increase, much of the plant growth dies and sinks to the
bottom to rot. Oxygen-using bacteria multiply as their
food increases and take oxygen from the water at a much
greater rate than aeration or plants can supply. Meanwhile a scum of rotted vegetation rises to the surface, cutting down the amount of sunlight penetrating the water,
and this cuts down photosynthesis so that less oxygen is
released. All these factors reinforce each other, and the
result is a monstrous growth of live algae and scum which
chokes the rivers and lakes while the fish die for lack of
oxygen. The drainage of nitrogen from fertilizer has destroyed the self-purifying capacity of every river in Illinois.

Another great and increasing source of nitrogen pollution in water comes from the automobile. Nitrogen oxides
from auto exhausts go into the air and are precipitated in
rain which finds its way through the soil to rivers and
lakes. Modern improvements to high-compression engines
have aggravated this problem, because they operate at
higher temperatures than older models, and at high temperatures the oxygen and nitrogen in the air combine more
rapidly to form nitrogen oxides. It is estimated that automobile exhausts contribute an amount of nitrogen in water
which is about one-third that contributed by fertilizers.

A certain unknown amount of nitrogen and phosphates
is contributed by industry, particularly where bleaching
and detergents are used, but since about 40 per cent of industrial waste is disposed of through municipal sewers, we
will go on to discuss the problem of sewage disposal and its
impact on oxygen depletion.

Until about a hundred years ago in the United States,
sewage was dumped directly into streams and lakes. Today, some 1400 communities, including good-sized cities

like Memphis, still dump their sewage untreated into the waterways. However, about 90 per cent of the population is served by sewage-disposal plants of which there are three general types: primary, secondary, and tertiary, depending on whether they have one, two, or three stages in the processing of the sewage effluent.

In one-stage plants the treatment is purely mechanical. As the waste-laden water, or effluent, enters the plant, it is screened to remove large floating objects and then passes through settling chambers where the heavier filth settles to the bottom as raw sludge. The effluent is then discharged into a river, lake, or coastal waters.

The process removes about one third of the gross pollutants, but this is not enough so that the waters can purify themselves naturally through their bacteria. The remaining two-thirds, mostly organic material, overwhelm the oxygen-using bacteria by a fascinating built-in contradiction. It provides so much food that the bacteria multiply at a fantastic rate. One would expect the great numbers of bacteria to take care of the increased organic material, and they do, but . . . the more the bacteria work, the more oxygen they use up until they reach a point where they use more than can be replaced, and then the bacteria die.

The obvious solution is to give the bacteria more oxygen. This is exactly what is done when a secondary stage is added. Oxygen-using bacteria are used *inside* the plant under controlled conditions and supplied with air. What nature does slowly outside the plant, man causes to be done rapidly inside the plant. After the settling chambers, the effluent flows onto a "trickling filter," which is a bed of rocks several feet deep. The rocks are covered with bacteria (normally present in sewage) and the bacteria multiply and decompose the organic matter. In some plants an aeration tank is used instead of a trickling filter. The effluent is mixed with air, saturated with bacteria, and allowed to remain for a few hours. Properly executed, either variety of secondary treatment will reduce organic matter by 90 per cent.

The only trouble with this type of plant is that in solving one problem it creates another. It does not remove phosphorus and nitrogen from the effluent but on the contrary, its bacteria turn the organic forms of these nutrients into mineral forms which are more usable by algae and other aquatic plants. The result, as we have seen, is to recreate the oxygen-consuming organic material that the sewage plant was supposed to control in the first place.

Some 60 per cent of the population is served by the two-stage plants and 30 per cent by one-stage plants. The present goal of the antipollution program is to extend secondary treatment to all the primary plants and most industrial plants at a cost of 13.5 billion dollars (10 from public funds and 3.5 from industry), but it is clear that sewage-disposal technology must improve. One scientist has attacked the multibillion-dollar program as an insane waste since it reproduces the problem. While other scientists do not go this far, they all agree that something has to be done about sewage plants. One answer is to take the effluent through a tertiary stage, such as that embodied in a modern plant at Lake Tahoe in California. As the diagram on the following page illustrates, the effluent, after aeration in the secondary stage, passes through a series of installations which remove phosphorus and nitrogen, and finally is filtered through charcoal. The result is water which is nearly fit to drink and is excellent for irrigation. A holding

PRIMARY TREATMENT

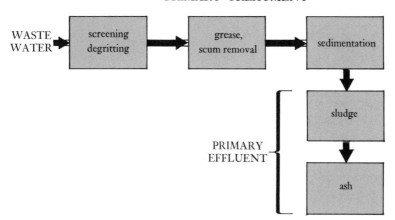

SECONDARY TREATMENT

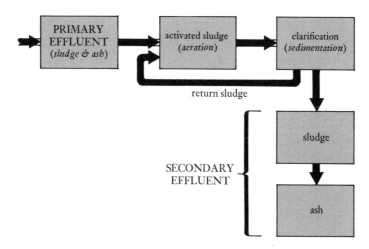

ADVANCED TREATMENT

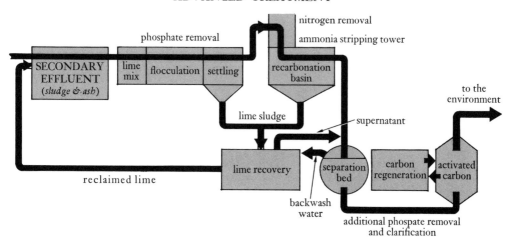

pond of nearly a billion gallons is part of the installations at Lake Tahoe. Late in 1970 the pond's waters were so clean that it was opened to the public for swimming, and when it is stocked, for fishing. Even more dramatic are the results of an extended tertiary treatment in Windhoek, South Africa, where the effluent is so clean that it is put back into the drinking-water mains—the first such instance in history.

We have been emphasizing so far the problem of oxygen depletion by phosphates and nitrogen only because this problem is the largest and most acute. But there is also the problem of persistent pesticides and herbicides. Because they affect the air as well as the waters and because they form a very large subject, we shall treat them in a separate chapter.

In addition, there are literally hundreds of pollutants—dyes, acids, etc.—produced by modern industry and agriculture which are not affected by bacteria. The tertiary treatment can deal with some of these pollutants but not all, and new technologies must be developed. Moreover, tertiary treatment is expensive, 30 cents per 1,000 gallons compared to 12 cents per 1,000 in two-stage plants. Finally, the tertiary treatment doesn't solve the problem of sludge disposal, which is becoming acute in major cities. The heavy sludge which settles at the bottom of the tanks is compressed, dried, and either burned (contributing to air pollution), or buried. Chicago has to dispose of 1,000 tons of sludge a day. It burns some, but buries about half of it at a cost of $60 a ton.

The remedies for water pollution are as many as there are sources of pollution—more, for any one source can be tackled from many angles. Thus to get rid of phosphate pollution we can improve the treatment of sewage *and* cut down on fertilizers *and* take phosphates out of detergents (which is being done). In general, it is more effective to attack a single source from many angles because it is rare that the pollutant can be banned by edict, such as was done with DDT. One cannot ban automobiles, or fertilizers, or power plants, or domestic sewage.

In all cases of pollution, without exception, there is one remedy that is readily available: cut down on the polluting activity. We can drive less; we can have fewer automobiles. We can cut down on the size of our newspapers, on the elaborate packaging of foods and swaddling of goods in paper and cardboard. We use four pounds of paper *a day* per family; England, Sweden, France, Japan—every bit as literate as we are and every bit as finicky about cleanliness and wrapping—use a fraction of that amount. This would cut down on the pollution of pulp mills, the air pollution of burning the paper, and save some forests in the bargain.

We can cut down on the extravagant use of electricity, the excessive use of lights. The same goes for radio and television sets which are often left on for hours on end when no one is using them. We can cut down on the electric gadgets; the electric knives and electric toothbrushes and electric vibrators and electric shoeshiners and electric can openers—on and on, the whole litany of so-called time savers whose only certain contribution to our society is the frustration engendered when they break down. Everyone can make his own list, including less overheating in apartment houses. Any saving in energy is a saving in water and air pollution.

Even where the waste is clean—for example, water

leaking from pipes or faucets directly into sewer systems—there is a price paid in pollution somewhere. It has been estimated that in the United States there is a per capita waste of 30 gallons of water a day—all of it clean, all of it expensive. In a town of a hundred thousand people, this comes to three million gallons a day. This means power was used, and power means pollution. Moreover the extra "clean" water adds to the load of sewage plants and may even prevent them from functioning properly.

A startling example of this "clean" pollution is the problem of storm waters in cities. Storm waters (water from rain and snow) are fairly clean, but they run off the streets through the regular sewers that flow into the treatment plants. Since storms add sudden floods of waters during heavy rains, practically all sewer systems have interceptor channels which, in a heavy rain, divert some of the sewage directly into streams to prevent the sewage plants from flooding. An important remedy, therefore, is to separate storm sewers from sanitary sewers in the 13,000 communities which now have combined sewers. The cost is a staggering 48 billion dollars. An alternate method, less effective, is to build holding tanks for the overflow storm water. This would cost 15 billion dollars.

An important contribution to reducing water pollution is industrial recycling of water. Recycling is the technical name given to the process of reusing waste water in the manufacturing process after the water has been purified sufficiently so that it can fulfill its function in that process. Such recycling costs money. Industrial leaders are as concerned about pollution as the rest of the community, but they are in business to make profits, and antipollution measures cost money. So industrialists tend to drag their heels. Moreover, as industry is constantly expanding, a reduction in pollution of, say, 10 per cent may be offset by a 10 per cent rise in production. New industrial processes are constantly developed, and some of them cause pollution in unforeseen ways. Even old and seemingly harmless processes may suddenly prove to be extremely dangerous. The most dramatic example in recent years is the discovery of mercury pollution. One third of all mercury is used in the chlorine industry, and another 17 per cent in the making of pesticides, paint, pulp and paper, in the treatment of seeds to prevent spoilage, and other processes. Fifty per cent is used in machines and industrial devices, in laboratories, in dentistry, and in medicine. From the chlorine industry alone an estimated 1.2 million pounds is released annually into American waters as industrial waste. A single plant such as the one of Dow Chemical Corporation on Lake St. Clair, which has been operating for 40 years, releases as much as 70,000 pounds of mercury a year in its waste.

Since mercury is considered an inert metallic element, no one worried about it until it was discovered that bacteria turn the inert metal into methyl mercury, which is poisonous to humans and which becomes concentrated in the flesh of fish. In March of 1970 the Canadian government banned all commercial and sport fishing on their side of Lake St. Clair because lethal doses of mercury had been found in fish. One month later, Americans found that the Detroit River and Lake Erie were full of poison, and four states, Michigan, Ohio, Vermont, and Alabama, banned all fishing in their rivers and lakes. By late 1970, 16 states had found high levels of mercury in fish, drinking water, and game birds, and had instituted various health restric-

tions. Thirty-three states have found contamination by
mercury, and some investigators are urging that the entire
Mississippi River be declared off-limits for fishing. In
July, 1970, the federal government instituted a whole
series of lawsuits to force companies to cut off pollution,
and most companies quickly complied. In a matter of a
few months, the chlorine industry reduced its mercury
waste by 95 per cent!

As agencies became alerted to the danger of mercury,
and as more sensitive detection instruments were devised,
the menace of mercury became more apparent. The Agriculture Department has forbidden the production and
interstate shipments of mercury-treated seeds, and on
October 29, 1970, the Food and Drug Administration announced the first recall of a product contaminated by
mercury—liver pills derived from seals killed on the
Pribilof Islands off the coast of Alaska. The producer of
the pill said that those animals had been specifically chosen
as the most free from contamination, adding ruefully:
"You can just figure from this that there isn't any place in
the whole earth that is not contaminated."

Late in 1970 it was discovered that tuna and swordfish
from many parts of the world were heavily contaminated
by mercury, and by early 1971 the Food and Drug Administration had ordered the recall of many brands of
canned tuna and nearly every brand of frozen swordfish.
The seafood industry was hard hit as customers throughout
the country became aware of the threat to their health and
stopped buying tuna and swordfish in stores and restaurants.
There were bitter jokes: one cartoon showed a counterman
shouting to the chef, "One tuna on rye—hold the mercury!"

As concern over water pollution grows across the country, industry is beginning to take remedial steps, either voluntarily or through the pressure of citizen groups and government agencies. The pulp and paper industry is one of the worst water polluters in the country, using up some 2.5 billion gallons of water a day. Two thirds of this water is discharged after inadequate treatment that does not meet federal standards and one third is discharged with no treatment at all. A citizens' group, the Council on Economic Priorities, issued a 400-page report (*New York Times,* December 17, 1970) that showed only two major companies, Weyerhaeuser and Owens-Illinois, as having adequate controls, established voluntarily over many years. The rest of the companies have proved recalcitrant. The report indicates a cost of 750 million dollars for the industry as a whole to develop adequate controls, but this is not too onerous when considering that the industry grosses 20 billion dollars annually and is the largest private landowner in the United States.

Because of popular concern, the paper companies are beginning to move. The St. Regis Company has planned capital outlays for pollution control of 70 million dollars in the next three to five years. A most impressive effort is being made by the Crown Zellerbach Corporation, owner of giant paper mills, which has just built a series of experimental lagoons at Lebanon, Oregon. There, organic wastes from papermaking are held, aerated, and bacteria is used under controlled conditions to produce an effluent with low-oxygen demand that can be discharged into a river. The project cost $800,000, jointly financed by the company and the government.

The steel industry, the oil industry, and the chemical

industry—all major polluters—are beginning to spend large sums for pollution prevention in ways similar to that of the pulp industry. As new plants go up, pollution controls are designed into the structures of the plants. Expenditures of such work rose from 2 per cent of capital outlays in 1967 to 4 per cent in 1968. In some companies, the increase has been greater: Bethlehem Steel is now at 6 per cent and will be increased to 10 per cent in the next five years. *Fortune* magazine reports that in two extreme cases, the figure is 30 per cent.

National estimates of the costs to fight water pollution (or any kind of pollution) are difficult to make. A joint survey (in 1970) by the National League of Cities and the United States Conference of Mayors estimates municipal needs at 33 to 37 billion dollars over the next six years. It is estimated that industry should spend about a third of what municipalities spend: add 12 billion. Storm sewer estimates are 48 billion, for a grand total of 90-100 billion dollars, or 10 billion dollars a year for 10 years.

Industry's expenditures on cleaning waste water have risen from 45 million dollars in 1952 to 600 million dollars in 1969. That's a big step forward, but still only a fraction of what needs to be spent annually over the next 10 years.

While voluntary efforts by industry are praiseworthy, they are sporadic, partial, and in some ways economically unfair to the individual companies concerned. The conscientious companies have their costs (and prices) go up, while the less responsible companies gain an advantage. Government standards would equalize the burden for all companies, and a mix of tax rebates and tax penalties would help enforce the standards.

An important antipollution measure is recycling waste—

cleaning the water and reusing it in the same process. This has been done very successfully in the Ruhr Valley in Germany, which has probably the greatest industrial complex in the world—steel mills, coal mines, chemical and pulp factories. A study released in April, 1970, shows what can be done. Cooling operations in blast furnaces have reduced the use of water by 10 times. A corrugated cardboard factory at Ebenhausen uses settling tanks to remove sludge and cooling towers to remove heat. The factory now renews its water supply *only twice a year*. Industry as a whole in West Germany now meets more than 60 per cent of its water requirements by recycling its waste.

Recycling water in industry not only treats waste, but conserves water. Conservation of water by itself cuts down on pollution of all types, because water is becoming scarce in various localities. Finding new sources and transporting the water via aqueducts, pumping stations, dams, etc. entails a great deal of industrial production, the use of electricity, construction activities, and so on, all of which add to the pollution of the environment. Therefore, water conservation is of prime importance, not only in industry, but in agriculture and in domestic use.

In agriculture, the largest saving potential is in irrigation systems. Most of the water in irrigation ditches evaporates and only a small percentage goes into the soil. Although all water that evaporates ultimately condenses somewhere as rain, snow, and sleet, so that the earth doesn't lose it, the condensation may take place in areas inaccessible to man. Cutting down evaporation would mean a great saving, and experiments are being made to cover irrigation ditches with plastic sheets.

Another area of experimentation is to put perforated

plastic pipes in ditches and cover them with earth. This is the so-called drip irrigation method which the state of Israel has been pushing for some years with notable success. It gets water to the roots of crops quicker than does sprinkler or ditch irrigation, with a marked increase in crop yield (usually more than twofold), a shorter growing season, and the obtaining of crops in areas of high soil salinity. These advantages offset the increased cost of irrigation, and there is a very great saving in water—perhaps as high as 75 per cent.

Conservation of domestic water may not seem important, if one considers that such consumption is roughly one tenth of total consumption (see table below), but one must remember that domestic water is very expensive, since it is for drinking and often has to be carried great distances. So even small savings are well worth the effort. Cutting back personal consumption to the 1950 level of 120 gallons would allow for a 25 per cent increase in population with no increase in water consumption.

This table shows the population of the U.S. (in millions) and water consumption (in billions of gallons daily). Domestic consumption is further broken down in daily per capita use.

YEAR	POPUL.	TOTAL	WATER	IRRIGA-TION	IN-DUSTRY	POWER	DOMESTIC Total	PER CAPITA
1900	75	40	(7)	20	(2)	10	5	5	70 gal.
1950	150	200	(35)	100	(20)	38	45	18	120 ”
1970	205	400	(75)	160	(45)	86	132	34	165 ”

(The figures in parentheses show amount that came from ground water by pumping)

The United States has been prodigal in the use of water. While the population doubled in 50 years since 1900, water consumption multiplied five times. The consumption doubled in the next 20 years with only a third increase in population. As discussed at the end of Chapter 1, by the 1980's we will be pushing against the limits of readily available water. Many areas are already in serious trouble and are pumping water from deeper and deeper aquifers which in some cases are being permanently depleted.

In Nebraska, for example, the water table has fallen 15 feet, and since most of it is irreplaceable fossil water, the supply will be exhausted in 50 years. In Texas, on the high plains, citizens are drawing seven million acre-feet of water from a water table that is recharging at the rate of 50,000 acre-feet per year. Tucson, Arizona is pumping water at rates far greater than replenishment rates. In the Mississippi basin the water table has fallen 400 feet in 30 years; in Los Angeles the water table is so low that the ocean is encroaching on it, and so on.

While conservation, recycling, and tertiary treatment plants all contribute to ameliorate the pollution problem, none of them fully answers the problem created by new and sometimes deadly chemicals in use at present or that may be developed in the future. Some of these chemicals, for example, kill the bacteria on which the secondary treatment depends. Also some viruses lethal to man go through unscathed under present techniques. It is imperative therefore that new techniques be developed for sewage treatment. One technique, now being studied, is to use chemical treatments to replace bacteria treatment—in effect, to do artificially what nature has been doing. Another technique being explored is called "reversed osmosis." Osmosis is the passage of liquids through a semipermeable membrane. If the two bodies of water on either side of the membrane have different concentrations of materials (say salt,) the less concentrated will flow naturally toward the more concentrated. Thus, if sewage is on one side of a membrane and clean water (in a lake, for example) on the other, the clean water will flow across the membrane to the waste. Reverse osmosis is exactly what it says: it reverses the natural flow by applying pressure on the waste side, forcing the waste water molecules to go toward the clean water. In the process, the membrane stops nearly all the pollutants. It's a promising process, except that the membranes devised so far get fouled up. Engineers are trying to solve that problem.

Ideally, however, the best way to treat waste is to return it to the soil as fertilizer, thus reestablishing the natural cycle. Earlier we gave the example of Holstein cows (p. 43) which were removed from pastures to feed lots. Their waste, which formerly had fertilized the pastures, was flushed into sewage systems or streams. It is impossible for practical reasons to get the cows back onto pastures: it would cost too much and cut down on both the number and weight of the animals. But it should be possible to get the effluent back from feed lots to the soil. This is no small matter, for there is ten times as much animal waste as there is human waste. In 1966, 10 million head of cattle were in reed lots, creating the waste equivalent of 100 million human beings (half the population of the U.S.).

To return waste to the soil is hardly a new idea. A century ago London was doing it, and both Paris and Berlin have had large scale projects for many decades. Municipalities in the United States have lagged behind, in part because of reluctance to invest in land outside their boundaries and in part because not enough study has been made of the problems involved. Under contemporary pressures, these attitudes are changing.

Increasingly, scientists are working on getting effluent back to the soil. Chicago has been conducting pilot projects. One of them has taken up the problem of sludge, which Chicago burns or buries. The dried sludge is compressed and shipped to Florida as fertilizer, but the process is still too expensive. Another pilot project is to pump the effluent through a pipeline to the farms and strip mines seventy miles from Chicago. This, too, gives promise. It has been estimated that the cost of such a pipeline, with pumping stations, reservoirs, irrigation equipment, and manpower will come to 20 dollars a ton. Since Chicago is paying 60 dollars a ton to bury sludge, it would pay the city to make the shift. In fact, the city could build a pipeline some 200 miles long and still be within the present cost of disposal. The problems are many, and studies now in progress are seeking solutions. One of these studies, the results of a five-year experiment at Pennsylvania State University, was published in March, 1970, and makes exciting reading.

Pennsylvania State University and the nearby town of State College are located in the Nittany Valley, near the center of the state. The Nittany Valley is drained by Spring Creek, which is joined by Slab Cabin Run, flowing into Bald Eagle Creek, and thence into the Susquehanna River. Spring Creek is fed by many tributaries and springs which are fed from the underground waters. The University has been pumping out of the creek some two million gallons of water a day, and the water table dropped 60 feet in three years. A two-stage sewage plant emptied its waste into Thompson Run, which is a tributary of Slab

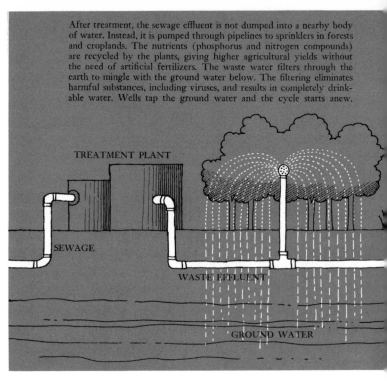

After treatment, the sewage effluent is not dumped into a nearby body of water. Instead, it is pumped through pipelines to sprinklers in forests and croplands. The nutrients (phosphorus and nitrogen compounds) are recycled by the plants, giving higher agricultural yields without the need of artificial fertilizers. The waste water filters through the earth to mingle with the ground water below. The filtering eliminates harmful substances, including viruses, and results in completely drinkable water. Wells tap the ground water and the cycle starts anew.

TREATMENT PLANT

SEWAGE

WASTE EFFLUENT

GROUND WATER

Cabin Run. Tests showed that there was one and a half times more nitrogen and 50 times as much phosphorus in the stream after it had received the sewage. Algae began to grow until they choked Spring Creek, and the once plentiful trout disappeared.

An interdisciplinary team at the University tackled the problem in the early 1960's. The team consisted of agricultural and civil engineers, agronomists, biochemists, foresters, geologists, microbiologists, and zoologists. They studied the type of soil, the depth of the water table, the depth of soil layer between surface and bedrock, and many other factors. They decided to return the effluent to the land in a series of carefully controlled experiments in both forest areas and cropland. A pumping station was added to the sewage plant, and the effluent from the plant was carried to the experimental plots. For each plot irrigated by effluent

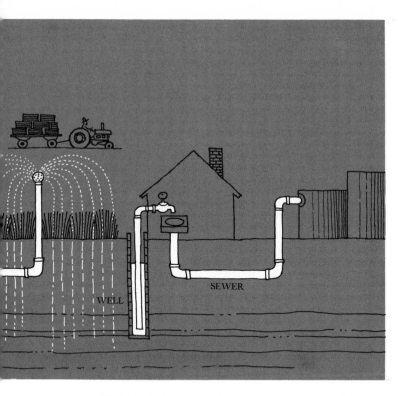

(which had been analyzed as to composition), a similar control plot was irrigated by well water and artificial fertilizer, by measured amounts, once a week from April through November, for five years. The effluent was not drinkable, and a purpose of the experiment was to determine whether the soil would filter the water enough to make it drinkable before it joined the underground water. The experiment began on May 16, 1963.

The results were published in the magazine *Environment* of March, 1970. They show that the plots irrigated with effluent outproduced the plots with artificial fertilizer by considerable margins, which varied with the crop—in the case of alfalfa, by more than two and one-half times. They show that young oak trees had an annual increase in diameter from 27 to 83 per cent greater than those in the control areas.

Most important, however, was the proof that soil as a "living filter" completely purified the waste water before it reached the water table. The soil filtered out phosphates, nitrogen, mineral salts, detergents, bacteria, and particularly coliform bacteria. While coliform bacteria are themselves not disease bacteria, they live in intestines and their presence is the best index of contamination by fecal matter. A coliform count of 2,400 per 100 millimeters of water is considered the criterion of whether or not water is safe for swimming. Samples taken at two-foot depths showed only one coliform bacteria per 100 millimeters of water.

Perhaps the most important health consideration is the fact that the soil eliminates not only coliform bacteria but viruses. There is a growing suspicion among public health physicians that many rapidly spreading diseases are transmitted by viruses. Sequestration of effluent in oxidation ponds kills about 70 per cent of viruses within a month, but only filtering through soil causes the viruses to disappear altogether. Soil particles have an electrical affinity for viruses which allows the particles to hold viruses long enough to be dismembered into innocuous proteins.

The experiments at Pennsylvania State marked a significant breakthrough in sewage disposal, but of course it was too small a test to be conclusive. Normally it takes decades for a new process to move from experimentation to full-fledged commercial projects, but by a quirk of good fortune, a major project was launched within a year in Muskegon County, Michigan, embracing 12 cities and townships with a population of 170,000. In September, 1970, the Federal Water Quality Administration announced a grant of one million dollars for research and demonstration plus another million dollars for construction. These sums cover the first year of a seven-year commitment. The total cost is estimated at 30 million dollars, of which 55 per cent will be federal money, 25 per cent Michigan State money, and 20 per cent Muskegon County money.

The disposal system has been planned by the Bauer Engineering Company of Chicago, and contracts were let on January 1, 1971, with a goal of beginning operations in 1972. The Bauer plan is bold in scope and farsighted in conception. It envisages that all the sewage in the area will flow into a huge outlet pipe that swings away from Lake Michigan, Muskegon Lake, Mona Lake, and White Lake —the traditional sinks for the community wastes—into a barren sandy area fifteen miles away where the pipe will empty into three aerating lagoons of eight acres each.

Held for three days under continual aeration by mechanical mixers, bacteria would decompose the waste matter and then the effluent would move into two storage lagoons of 900 acres each. The purpose of those reservoirs would be to hold the effluent during the winter months when the frozen ground cannot absorb the liquid. In the spring, the effluent would be piped out to rotary irrigation rigs which would spray the liquid and its suspended particles over 6,000 acres of potentially fertile land. It is estimated that there may be a profit of as much as $750,000 from the sale of corn grown on the irrigated acres. Moreover, it has been suggested that a new industrial complex to utilize the corn might be built in the area, providing jobs in a rather poor area. Finally there is the advantage of clean shorelines on the now polluted lakes, with all the potentials of recreation—fishing, swimming, and so on.

If this system works (and the chances of success are very high), it can be applied to nearly all the major municipal areas in the United States and constitute a boon to mankind of incalculable value, not only in solving the water pollution problem, but also in solving a major problem in agriculture—the excessive and dangerous use of nitrogen fertilizers.*

Fertilizers are now used in enormous quantities that have increased 14 times in the last 14 years. A large proportion of these fertilizers is not used up by crops but drains off the land through rainfall and goes into the water supply both on the surface and underground. Excess of nitrogen in drinking water causes the serious infant disease of methemoglobinemia, and already in California and the Midwest the level of nitrates in many areas has been raised above the safe limit recommended by health authorities.

One of the greatest dangers of the increased use of nitrogen fertilizers is that, after a time, many species of nitrogen-fixing bacteria may become extinct. Ecologists are beginning to be aware of this possibility, and it is a worrying one because, if this should happen, mankind could never go back to a natural ecological system. In an irreversible setup, farmers would be forced to use synthetic fertilizers for evermore—an ecological disaster of the first magnitude. Success at Muskegon would fire the imagination of even the stodgiest public official and would provide a tremendous fillip to the drive to clean up *all* water pollution in the U.S. within a decade—a dream to beat any LSD trip.

---

* The fascinating tale of the politics behind this success story is told by John R. Sheaffer in the *Saturday Review*, November 7, 1970. Sheaffer, while at the Northeast Illinois Planning Commission, was in touch with the experimenters at Pennsylvania State and was instrumental in getting Bauer to apply the results. His contribution has been recognized by President Nixon in a personal letter, and he is now with the U.S. Corps of Engineers.

● 5 ●

OCEANIC BLUES

Wherein we find that the mighty oceans are not immune to pollution. The importance of coastal waters and their poisoning. How all life in the deep oceans becomes affected. The danger of oil pollution and some suggested remedies.

Until the late 1960's, the oceans were considered impervious to pollution. The amount of water in the oceans is so vast that people in general and sanitary engineers in particular thought the oceans inexhaustible in their capacity to absorb all the garbage and sewage man could dump in them. As recently as 1966 a sanitary engineer from Los Angeles waxed lyrical on the subject:

> The great economy inherent in the discharge of urban sewage and industrial wastes into near-shore waters for final disposal is apparent . . . If the ocean, or one of its arms, can be reached with a sewer outfall, the grim specter of an expensive complete treatment plant grows dimmer and dimmer until it fades entirely . . . the good old ocean does the job free.
> And small wonder that we look to the sea for this assist. Its vast area and volume, its oxygen-laden waters, its lack of potability or usefulness for domestic and most industrial purposes, present an unlimited and most attractive reservoir for waste assimilation.

In the light of information readily available, the remarks can only be described as fatuous. The oceans can be and are being polluted. The oceanic vastness is deceptive since the surface waters tend to be stable. Because warm water stays on top, and the sun warms the surface, the turnover from the oceans' depths is very very slow, and the mingling of the various strata takes many centuries. Vast amounts of solid waste float on the surface, taken by winds and currents to the remotest parts of the oceans. Beer cans have drifted on the shores of the Galapagos Islands, a remote and uninhabited spot in the South Pacific. Thor Heyerdahl, of Kon-Tiki fame, led two expeditions across the Atlantic, in 1950 and 1970. He was horrified at the pollution built up in the twenty-year interval.

Ocean pollution ranks high among the problems of global pollution. In the summer of 1970 a conference of leading American scientists and administrators took place in Williamstown, Massachusetts, under the auspices of the Massachusetts Institute of Technology, for the specific purpose of studying global environmental problems. One of its four work groups devoted about half of its sessions to oceanic pollution.*

As could be expected, the highest degree of pollution is found in shallow waters along the coasts and in the lowlands where land and oceans meet—the marshes, the estuaries of rivers, the beaches. The so-called estuarine zone was once treated with little consideration: estuaries were dredged, marshes filled, the offshore waters of large harbors used as dumping areas for garbage. For example, for forty years, New York City has taken its garbage in barges five miles out to sea, and there dumped. Suddenly the city has learned that it has created a 20-square-mile area of "dead sea" where pollution is so high that fins and scales of fish rot off their bodies.

---

* A report of the conference has been published by the MIT Press, *Man's Impact on the Global Environment*, Cambridge, 1970, and has been referred to in this book.

It now turns out that the estuarine zone is crucial to life in the sea. A leading authority on marine ecology, Professor William A. Nearing of Connecticut College, has estimated that 90 per cent of our total seafood harvest is dependent in one way or another on the estuarine zone and this zone is being polluted beyond imagination by sewage, garbage, trash, oil, phosphates, nitrates, pesticides, and other long-lasting chemicals. The impact on coastal waters and wetlands can be seen in one single example: along the Georgia coast the harvest of oysters fell from 8 million pounds in 1908 to 160,000 pounds in 1960!

The Department of the Interior has classified some 20 different kinds of wetlands in the United States. Originally they comprised 127 million acres, but now only 70 million (about 60 per cent) are left. Of these, 27 million acres distributed among 27 coastal states were regarded as important to fish and wildlife, and already some 7 per cent (one and a half million acres) have been wiped out. California leads in the loss with 250,000 acres out of 380,000 acres being eradicated. Northeastern Florida is a close second. New York, New Jersey, Connecticut, and New Hampshire have lost between 10 and 15 per cent of their estuarine zone to filling, dredging, etc. In the remaining acreage, pollution is on the increase as the coastal states increase in population (and consequent sewage) and in industry. Two thirds of all pesticide factories, two thirds of all factories turning out organic chemical products, 60 per cent of those turning out inorganic chemical products, 50 per cent of all refineries, and two thirds of all pulp mills are located in the coastal states. Pollution of beaches near the large towns is steadily on the increase, to the discomfort and discontent of the people.

The shallow coastal waters are particularly rich in planktonic life and in the small fish on which the life of the sea
depends. Pesticides and other chemicals are taken up by
these organisms and passed up the food chain as the bigger
fish eat the smaller, until heavy concentrations can be
found in good-sized fish such as the kingfish. For example,
on January 1, 1971, federal agents revealed that they had
impounded four tons of frozen kingfish caught during
December about 20 miles off Los Angeles because the fish
showed 19 parts per million of DDT, which was four times
the federal limit allowed for humans. Three parts per
million is lethal for tiny fish, and at 25 parts per million
humans begin to be affected.

With these figures in mind, consider the astonishment
of scientists when 60 dead sea lions were washed ashore on
the beaches of California in October of 1970 and an examination showed 4,000 parts of DDT per million! The same
sea lions also showed concentrations of mercury. The seals
whose livers were made into poisoned pills, as described in
the previous chapter, showed concentrations 116 times
higher than is considered safe in humans.

Sea lions, seals, and tuna are large species, and they are
deep sea animals which do not usually come close to land.
The seals, for example, weigh up to 600 pounds and range
from 100 to 1,000 miles off the coastline. The poisoned
food chain that starts with tiny fish in the bays and passed
up through the larger fish offshore delivers death into the
ocean deeps. No part of the ocean is safe from contamination. Sperm whales have been found with concentrations
of DDT of six parts per million in their blubber and they
feed on plankton only in the far reaches of the oceans.

Fish-eating birds are vulnerable. Petrels have been found

to carry 10 parts per million of DDT. While such concentrations are not lethal to the birds, they affect their reproduction. DDT and its residues affect certain enzymes which control the thickness of eggshells and cause them to underdevelop as much as 20 per cent. This makes the shells so thin that they break during incubation. In a study in southern California, only one brown pelican was known to hatch out of 1,200 nestings in 1969. The bald eagle and the peregrine falcon have disappeared completely from the wilderness area of the Channel Islands. Many marine species are in danger of extinction, and once a species is gone, no power on earth can recreate it.

After reviewing the research of the last few years on the marine effects of DDT and its relatives, the Williamstown work group concluded: "In view of the findings of the past decade our prediction of the hazards may be vastly underestimated . . . We recommend a drastic reduction in the use of DDT substance as soon as possible."

We mentioned that the plankton diet of sperm whales was contaminated, and this was quite a puzzle to scientists. It was easy to see how the poisoned food chain reached far into the ocean, but how could plankton thousands of miles away from coastlines be contaminated with DDT? There isn't enough DDT to contaminate all the water in the seas; the contamination takes place in the shallow coastal waters and gets progressively diluted to infinitesimal amounts. The answer seems to lie with another polluting villain—oil. Oil forms a fine film on the surface of the waters, spreading far and wide, and carries DDT in solution. Since most of the oil pollution takes place in harbors, refineries, etc., in coastal areas, when coastal waters are heavily polluted with DDT, the oil picks up the chemical and, by means of tides, carries it all over the world.

The pollution of petroleum products is a major, if not *the* major, component of oceanic pollution. Petroleum is a very complex structure with some extremely poisonous or unpalatable substances which have long harmfully affected the food chain of sea life. As early as 1928 a few naturalists observed the effect of oil pollution on the birds. Wrote Henry Best: "A new danger threatens the birds at sea. An irreducible residue of crude oil, called by refiners 'slop,' remains in stills after oil distillation, and this is pumped into southbound tankers and emptied far offshore. This wretched pollution floats over large areas, and the birds alight in it and get it on their feathers. They inevitably die."

Today it is estimated that, exclusive of production accidents, about a million tons of oil is spilled annually into the oceans by leakages, flushing of bilges and tankers, and accidents to vessels. When accidents from oil wells (such as the spill in the Santa Barbara channel) and the industrial dumping in harbors are included, the total oil pollution in the oceans may be anywhere from 10 million to 100 million tons annually, according to Dr. Max Blumer, senior scientist at the Woods Hole Oceanographic Institute. While dragging a net to skim ocean life in the vast Sargasso Sea, Dr. Blumer found the oil film so thick for a stretch of 500 miles that the net had to be cleaned with solvent every two to four hours, and finally towing had to be suspended. (Max Blumer, *Oil on the Sea*, Plenum Press, 1969.)

The amount of oil pollution has risen sharply in the last few decades, due to a combination of factors: the steadily increasing number of offshore wells, the number of tankers operating, and the size of the tankers. All these factors will be on the increase in the near future. The size of tankers is particularly worrisome: the *Torrey Castle*, which broke up on a reef off England, spilled a cargo of 100,000 tons of oil; supertankers of 300,000 tons are afloat—there were four giant tankers of 327,000 tons in service in 1970. Huge tankers of 500,000 tons are under construction, and tankers of 800,000 tons are being projected. A single spill from one of those supertankers would add as much as 20 per cent to the present yearly oil pollution.

While accidents can be reduced, they cannot be eliminated, and each accident tends to do more damage than in the past. At present, 1,000 million tons of oil a year is shipped on the seas, and the amount is constantly increasing. Submarine reservoirs of petroleum are likely to be found on the continental shelves of almost every continent, with a great increase in the kind of production accidents that took place in the Santa Barbara channel and in the Gulf areas. Local in nature, this pollution spreads out. Winds and currents concentrate it into such zones as the Sargasso Sea, where investigation shows that there are more oil globules and tar balls than the Sargassum weed. Heyerdahl described a continuous stretch of 1,400 miles so fouled with oil that his crew couldn't take a swim—right in the open Atlantic!

Ocean pollution is on the increase. Unless drastic measures are taken to cut down on pollution, as well as new techniques developed to neutralize oil spills, the oceans are bound to suffer. Since all the waters are interconnected, damage in one area will spread to all the rest. Scientists believe that a thin film of oil is spreading over all the oceans, restricting the sunlight that penetrates the surface to fuel photosynthesis. In each season, some 20,000 tons of vegetable matter is created for every acre of the ocean surface. A reduction of this harvest would undermine the marine food chains and drastically reduce the seafood so desperately sought by hungry man.

A source of contamination of the oceans that is difficult to evaluate is that of radioactive wastes. Amounts of radioactivity in the Irish Sea are high enough to cause embryo fish to develop deformed backbones. The pollution comes from the Windscale nuclear power station on the English coast. More nuclear stations will be built on coasts in the coming decades, as well as those being built on rivers that drain to the seas. Since many of the radioactive elements last a long time—decades and even centuries—the pollution of the oceans accumulates. The radioactivity in the oceans comes from three major sources: leakages and waste from nuclear plants, nuclear testing in the atmosphere,* and accidents to nuclear-powered vessels such as the sinking of the nuclear-powered submarine *U.S.S. Thresher* in 1963. Another possible source in the future may be the practice of dumping solid radioactive wastes in sealed containers.** No one really knows how safe these containers will be. An international conference of 260 scientists from 60 nations was held in the summer of 1970 on the island of Malta in the Mediterranean to study the problems of radioactivity and recommend action to the U.N.

A great deal more information is needed on this question of radioactive pollution of the oceans. The Williamstown

---

* The radioactive particles given off in tests drift with the air currents and are precipitated through rain all over the globe surface, including the oceans. Radioactive traces have been found in the snows of the Arctic. While there is now a treaty prohibiting tests in the atmosphere, neither France nor the People's Republic of China are members of the treaty, and they are continuing their tests.
** A similar worry applies to the sealed containers of nerve gas which have been dumped in the oceans.

Conference in 1970 excluded the subject from its delibera-
tions because of the paucity of information and recommended large scale studies and research.

There is bound to be an increase in oceanic pollution in
the next few years as nuclear plants and oil consumption
increase. Oil pollution is expected to quadruple by 1980 if
counteracting measures are not taken. Such measures are
being taken, but it takes time to get results. However, the
future is not as bleak as it appears. The ocean *does* have
enormous recuperative powers, and we are a long way from
a point of no return in pollution. All nations are concerned,
and it is probable that international treaties will be negoti-
ated to diminish ocean pollution in various categories. On
November 1, 1970, the United States proposed to the 14
nations which comprise NATO (North Atlantic Treaty
Organization) an agreement for drastic control of oil spills,
leakages, and flushing out of tankers. When adopted, this
may become a pattern for a worldwide treaty.

More important than the treaties is the fact that most of
the pollution takes place in coastal waters and hemmed-in
bodies of water such as the Baltic Sea, the Mediterranean,
the Black Sea, the North Sea, the Red Sea, the Persian
Gulf, the Sea of Japan, and so on. To stop pollution in
those areas is usually a national responsibility, or, in a few
cases, the responsibility of a few nations which can negotiate regional treaties. It is much simpler to negotiate among
four or five nations than among a hundred-odd.

Every nation has a direct and pressing interest to clean
up its coastal waters, since they are the most important
source of seafood, as well as recreation, tourism, etc.
Every step which reduces pollution in other areas—sewage, fertilizers, power plants, automobiles—helps reduce

coastal pollution. The shallowness of coastal waters, which increases the impact of pollution, also increases the potential of recuperation once pollution is stopped, because the ebb and flow of tides and coastal currents act as flushing agents.

The industrial nations which have the money and techniques to stop pollution are beginning to move at an ever-accelerating pace. By the time the underdeveloped countries qualify as major polluters, there will probably exist international agencies to protect the oceans.

Despite his technological achievements, modern man is still awed by the majesty of the ocean, humbled by its calm serenity as well as its titanic rage. A decent respect for our origins should help us keep it unsullied for future generations.

• 6 •

HALF-LIVES AND WHOLE DEATHS

Wherein we speak of thermonuclear energy for war and for peace, and the impact of radiation pollution. The expected increase of nuclear as well as conventional power plants and the problems of their waste. The problem of thermal pollution. A horrendous radiation hazard recently discovered.

On August 6, 1945, the United States Air Force dropped an atomic bomb on Hiroshima. In a matter of seconds, 160,000 people were dead or wounded, and many of the survivors were later to envy the dead. Somberly the atomic era was ushered in under the sponsorship of Mars, the God of War, and for many years it continued under this sponsorship as the U.S., Britain, the U.S.S.R., France, and finally China sought to build up their arsenals. Today the Soviet Union has enough atomic weapons to kill every man, woman, and child in the world three times over; the United States can kill every man, woman, and child in the world ten times over.

These arsenals of atomic weapons were built up over two decades, with constant improvement of lethal power, due to a long series of tests by all the nuclear powers, particularly by the U.S. and the U.S.S.R. During these tests, the world became slowly aware of the dangers of atomic radiation. Radioactive elements given off in the explosions were carried by air currents all over the world and settled on the surface, through gravity or rain. The term "radioactive fallout" became well known. Even though the fallout might be minute, it was very dangerous, because many radioactive elements have a tendency to accumulate in the organs of the body, and many are long-lasting. Their duration is estimated in so-called half-lives, that is, the length of time it takes for half a given quantity of an element to decay into another element. The element strontium-90, for example, has a half-life of 28 years. So if you start with, say, 10 milligrams of the element, after 28 years, five milligrams are still strontium and radioactive; after another 28 years, two and a half milligrams; after 84 years, one and a quarter milligrams, etc. In other words, after a hundred years, 10 per cent of the strontium is still radioactive. As a result of the testing that has already taken place, it is possible that everyone in the world, and almost certainly everyone in North America, the U.S.S.R., Europe, and Northern China has some strontium in their bodies. Strontium-90 concentrates in bones, and through mothers' milk, in infants' bones and teeth. These concentrations increased after each series of atomic tests.

In 1967 it was discovered that the "daughter" element of strontium-90, yttrium-90 (also radioactive with a half-life of 2.7 days), concentrates in certain organs such as the spleen, liver, pancreas, pituitary, thyroid, and the lung lymph glands, which are so important in controlling metabolism and resistance to infectious disease. These are precisely the organs in which cancer rates in Japan have increased since 1945 by as much as 600 per cent! The strontium-90 accumulates in the bones, creating a reservoir which constantly is decaying into yttrium-90. Yttrium-90 goes into the bloodstream and concentrates in these various glands, doing much more biological damage than strontium-90. A final consideration must be added, and that is genetic damage. Yttrium-90 also concentrates in the ovaries of women and the testes in men, thus resulting, it is believed, in fetal deaths and genetic changes. It is uncontested that radiation is more damaging to the foetus and to infants than to adults. The extent is controversial, and many of the arguments (and data) can be found in the *Bulletin of Atomic Scientists,* beginning in May of 1969 and proceeding through 1970.

How much damage has been done is not exactly known because research is still turning up effects of radiation since Hiroshima. Testing in the atmosphere has greatly diminished as a result of most nations having signed the Nuclear Test Ban Treaty of 1963 which prohibits testing in the air. Several nations, including China, France, and Israel, have refused to sign the treaty, but the reader should beware of passing a moral judgment on them. Since the two superpowers, U.S.A. and U.S.S.R., have done all the atmospheric testing they needed for their weapons, they can only demand that other nations not develop weapons if they themselves would destroy their enormous arsenals. Moreover, the treaty does not prohibit underground testing.

With atmospheric testing greatly reduced, the main dangers of radiation are from underground tests for both military and peaceful purposes, from radioactive wastes in nuclear production and nuclear power plants, and from accidents in the production process as well as from existing weapons.

The danger from accidents is difficult to assess because in the United States, the two agencies involved, the Pentagon and the Atomic Energy Commission have been, and are, less than candid toward the public. The Pentagon has control over all weapons and the AEC has control over all production whether for military or peaceful uses, and they have consistently followed a policy of maintaining as much secrecy as possible. While some of the secrecy is due to the needs of national security (other nations are equally secretive), a very large element is the desire of those agencies to keep the public off their backs. Both agencies want nuclear production, that's their mandate, and they fear that a concerned public will demand safety margins which will slow down nuclear experimentation, production, and use.

Occasionally, an accident cannot be concealed and helps to draw aside the curtain of secrecy. On the question of weapons, two accidents were particularly important; one in 1961 near Goldsboro, North Carolina, and the other in 1966 near Palomares, Spain. In the first, a 24-megaton bomb (i.e., equivalent to 24 million tons of dynamite) was jettisoned from a B-52 and fell in an open field. There are six interlocking switches which must be triggered in sequence to explode the bomb. Investigation showed that *five of the six* switches had been activated by the fall. In the second case, a B-52 crashed after colliding with a KC-153 refueling tanker in midair near Palomares, Spain, while carrying four H-bombs. Fortunately the bombs were "unarmed," that is a piece of mechanism was missing, but of course they held radioactive materials. One bomb landed intact in a dry riverbed, one sank in the ocean and was recovered after a three-month intensive search. Two broke upon impact and scattered radioactive material on populated farmlands. Contaminated soil was loaded into 4,810 steel barrels and shipped to the U.S., where it was buried at the AEC facility near Savannah, Georgia.

In all, the Pentagon has admitted to 33 major accidents between 1950 and 1968. Eleven took place between 1950 and 1960 and 22 since then. It has been estimated that there

are at least 50 more accidents in addition to those that have
been reported (*Environment*, July–August, 1970). Two of
the accidents involved two nuclear-tipped missiles which had
been accidentally launched, and Premier Khrushchev is
reliably reported to have told the then Vice-President Nixon
that an erratic Russian missile on its way to Alaska was
destroyed in midair.

It is practically impossible for outsiders to estimate the
amount of radioactivity involved in accidents with nuclear
weapons or nuclear-powered ships. We do know that in at
least three accidents the nuclear head was exploded. There
have been five instances of fire and/or explosions on aircraft carriers of types that carry nuclear weapons. Often,
no certain knowledge is possible, as in the case of the
nuclear-powered submarine *Thresher*, which went to the
bottom of the ocean in 1963. No one knows what is happening to its radioactive materials. The *Thresher* carried
Subroc, a weapon with nuclear warheads.

After the loss of the *Thresher*, Congressional hearings
were held in 1964 and brought out another area of hazards
in nuclear reactors, namely the lack of stringent standards
and quality controls in manufacturing. Admiral Rickover
testified at the hearings that:

> "During the recent stay of the *Thresher* at Portsmouth
> about 5 per cent of her silver-brazed joints were ultrasoni-
> cally inspected. These joints were in critical piping systems,
> 2-inch diameter or larger. The inspection revealed that about
> 10 per cent of those checked required repair or replace-
> ment. If the quality of the joints inspected were representa-
> tive of all the *Thresher's* silver-brazed joints this means that
> the ship had several hundred sub-standard joints when she
> last went to sea."

If such faults could be found in a warship subject to the

most rigorous and manifold inspections, what is the situation in commercial nuclear reactors? Admiral Rickover thinks it is not very good: higher and tighter standards must be set not only for newly designed components, but also for conventional components. After giving instance after instance of faulty valves and electronic components, shoddy workmanship in brazing, and so on, Admiral Rickover continued: "It is not well enough understood that conventional components in advanced [i.e., nuclear] systems must necessarily meet higher standards. Yet it should be obvious that failures that would be trivial if they occurred in a conventional application will have serious consequences in a nuclear plant because here radioactivity is involved." ("The Lesson of the *Thresher*" from the *Perils of the Peaceful Atom,* Doubleday & Co., 1969.)

The setting of standards and of quality controls in the manufacturing of nuclear reactors is the province of the Atomic Energy Commission. As of 1967, the AEC admitted that out of 2,800 to 5,000 standards necessary for a typical reactor power plant in the areas of materials and testing, design, electrical gear and instrumentation, plant equipment and processes, *only about 100 recognized reactor standards had been approved as of March, 1967!* Yet more than a dozen nuclear plants were in commercial operation!

Accidents have occurred as well in areas under direct control of the AEC. Of some 200 underground tests by the AEC, seventeen have vented seriously, that is, radioactive gases have escaped. In one instance some 6,000 sheep downwind from the test were killed or affected. How many accidents have happened in plants is not known, but it is known that in one plant near Denver, the Rocky Flats plutonium plant, there have been at least three fires, one small explosion, and one period of plutonium contamination from leaky drums. Plutonium contamination happens to be extremely serious for two reasons: a very tiny amount causes cancer of the lungs, and plutonium has a half-life of 24,000 years—for practical purposes it is permanent contamination. At Rocky Flats the AEC at first denied contamination, saying that "there is no evidence that plutonium was carried beyond plant boundaries" (press release, December 1, 1969), but had to retract when a group of scientists organized the Colorado Committee for Environmental Information and conducted its own tests. It was shown that discernible amounts of plutonium were in the soil outside the plant boundaries—some 100 to 1,000 more plutonium than there should be by AEC standards.

The Committee's work forced the AEC to tighten safety-monitoring procedures and to make studies now in progress of the amount of plutonium contamination near the plant and the degree of hazard it represents to people.

The Colorado Committee, through the American Civil Liberties Union, also brought suit against the AEC on another matter and won a landmark decision that the AEC can be sued by a private party and that the court, and hence the public, must receive data from the AEC on tests made under its Plowshare program. The events that led to the decision are worth relating both because they indicate a source of dangerous pollution, and because they show how the AEC operates as a law unto itself.

Plowshare program is the enticing title which the AEC has given to its testing for peaceful uses of nuclear energy. All kinds of ideas have been put forward, including the use of nuclear explosives to make harbors and cut canals such as the proposed sea-level canal in Panama. (Incidentally, in November, 1970, a Presidential Commission on this canal recommended that conventional explosives should be used.) One of the applications suggested was in the extraction of natural gas where explosions are set off deep underground to release the gas and to build up pressure, forcing the gas to the surface. The AEC decided to try a nuclear explosion for this purpose, and in 1966 worked together with the El Paso Natural Gas Company on Project Gasbuggy, which was a test detonation of a 26-kiloton nuclear device (a kiloton is the equivalent of 1,000 tons of TNT) in a sealed well in a remote part of New Mexico. A few Indians living nearby protested the test, but the AEC stressed the fact that there would be no radioactivity released at the time of the blast. What the AEC did *not* stress was that the well would be opened up a month later and a great deal of radioactivity deliberately released into the air. The AEC knew that the natural gas would become highly radioactive and planned to burn it at the head of the well, a process known as flaring. The burning would chemically change the gas and certain elements such as tritium (radioactive hydrogen) into water, itself radioactive, which would find its way into the subsoil and the water table. One dangerous element, krypton-85, which is chemically inert, would be released into the atmosphere. Some limited measurements made during the flaring by the Public Health Service showed that radioactivity in vegetation downwind from Gasbuggy had increased tenfold.

At this point one may wonder: why produce radioactive natural gas if it has to be burned at the head of the well? The answer is that the AEC and the gas companies intend to mix the radioactive natural gas with ordinary natural gas to dilute the radioactivity.

Project Gasbuggy was a single-shot experiment. The next project called Project Rulison was to consist of a shot twice as large as Gasbuggy, to be the first of a series of hundreds of shots to develop a 60,000-acre gas field. Fears were expressed not only because of radioactivity, but also because of seismic disturbances (earthquakes). AEC dismissed the dangers, but was proved wrong and had to pay over $76,000 in property damages when several earth tremors followed the Rulison shot. On the issue of radioactivity, however, the Colorado Committee sued the AEC through the American Civil Liberties Union to prevent the shot, basing its suit on the effects of Gasbuggy. The court's decision was a compromise on the shot: the AEC was allowed to set off the explosion on September 10, 1969, but the well could not be opened and the gas flared for six months after the shot (compared to one month in Gasbuggy). This would give time for some radioactive isotopes such as iodine-131 to decay to safe levels.

More important as a precedent, the court ruled that when the flaring takes place, the AEC must inform the court (hence the public), within thirty days, of all data pertinent to the release of radioactive gases, and that it would maintain jurisdiction over subsequent complaints regarding public safety. Furthermore, the court's permission was for one shot; the rest of the series is open to litigation.

What is significant about these incidents is that it was the public and the courts which forced AEC to take safety steps which the Commission should have taken on its own. The AEC attitude to health hazards from radiation is a very cavalier one, and even now Colorado is suffering from lack of standards set in the past. For example, the Atomic Energy Commission set no standards of ventilation for uranium mines when they began to operate 20 years ago. The result is that uranium miners have an incidence of cancer a hundred times that of the general population. Tailings (mine waste) from uranium mills are piled up and rain leaches radium-226 from these tailings into the surrounding streams to dangerously high concentrations. The AEC permitted 300,000 tons of these tailings to be used as a sandy base for foundations of some 3,000 buildings in the Grand Junction areas in Colorado, including two schools. Radon, a radioactive gas resulting from the decay of radium-226, seeps through the cement slabs of the foundation. In two homes the level was so high that in January, 1970, the families were evacuated.

A further weakness of AEC safety standards must be noted. Permissible levels of radioactivity are designed only for *human beings*, with no reference to aquatic life or other organisms. But many radionuclides are absorbed from water by both plants and fish and are passed on in the food chains, sometimes with a cumulative effect. Scientists from the Department of Conservation at Cornell University have pointed out that a proposed nuclear station on Lake Cayuga could be discharging lethal levels of strontium-90 in the lake and within a few years those levels could build up to dangerous doses in water, in fish, and in milk from dairy farms using the water.

Moreover, fish in their early stages are extremely sensitive to radiation. The Cornell scientists have come to the conclusion that nuclear power plants should be required by law to discharge *no radioactive wastes of any kind* in streams, lakes, or coastal waters. They point out that technology exists to make this law possible and should be applied.

Potentially, the most dangerous radioactive hazard lies in the high-level radioactive wastes that result from the production of plutonium.* Plutonium does not exist in nature; it is a man-made element created from uranium. Uranium is the heaviest element found in nature (an atom of uranium is three and one-half times the weight of an atom of iron), and it is the splitting of the uranium atom which made possible the atomic bomb. To understand how plutonium is made, and the problems of atomic energy, we need to know a little more about atoms.

The structure of an atom can be visualized as something like the solar system, which has a heavy mass (the sun) at its center and smaller, lighter bodies (the planets) in orbit around it. In the atom there is a heavy central mass, the nucleus, with a varying number of electrons in orbit around it. Inside the nucleus are two kinds of particles: protons and neutrons. A proton and a neutron weigh the same, but the proton has a positive charge of electricity while the neutron has none. The number of protons is matched by the number of electrons (which carry a negative charge) so that, unless disturbed, the atom is electrically stable. The number of neutrons in the nucleus is always more than the number of protons, but it can vary: the uranium atom, which has 92 protons, may have 142 or 143 or 146 neutrons. These variants of an element (in uranium known as U-234, U-235, and U-238) are called isotopes: they all behave the same chemically.

Some isotopes are very rare; others very common. About 99.3 per cent of all uranium in nature is the isotope

---

* The term "high-level" refers to intensity of radioactivity in contrast to "low-level" wastes in the industrial use of nuclear energy.

U-238. U-235 runs about .7 per cent, and U-234 is barely traceable. Only U-235 splits if an extra neutron is shot into it; U-238 will *not* split even if it absorbs an extra neutron.

When U-235 gets an extra neutron and splits, it divides into two particles of other elements with 14 neutrons left over.* The splitting releases an enormous amount of energy. *One* pound of U-235 is the equivalent in energy of *20 million* pounds of dynamite! It is those 14 free neutrons which make the bomb possible. Some get lost, but some get absorbed by other nuclei of U-235, which in turn release neutrons which hit other nuclei of U-235, and so on—the so-called chain reaction.

The U-238 when hit by a neutron does not split. However, the new U-239 is unstable and emits an electron from one of its neutrons. But a neutron without an electron becomes a proton. Therefore a new element is born, baptized neptunium, with 93 protons. NP, which is also unstable and does not split, emits another electron, giving 94 protons, or a new element called plutonium. Plutonium is stable, but if hit with a neutron it behaves very much like U-235.

Plutonium is much cheaper to produce than U-235 for two reasons: it uses the plentiful U-238 as a raw material and it can be separated chemically from the residues of the raw materials. The separation of U-235 from U-238 cannot be done chemically since they are the same element, and the process of separation is a very expensive one. Therefore little pure U-235 is produced today. Plutonium is used for weapons; for nuclear reactors in power plants a mixture of U-235 and U-238 is used—what is called

---

* The elements are barium (56 protons and 82 neutrons) and krypton (36 protons and 47 neutrons). Total neutrons are 129, and since U-235 has 143 neutrons the difference is 14 neutrons.

PRINCIPAL PARTS OF AN ATOM

THE ELECTRON usually has a negative electrical charge. It circles about the nucleus of the atom. The distance from nucleus to the electrons is more than 10,000 times the diameter of the nucleus.

THE PROTON is found in the nucleus. It is about 2,000 times heavier than an electron and has a positive electrical charge.

THE NEUTRON is also found in the nucleus. It has the same weight as the proton but it has no electrical charge.

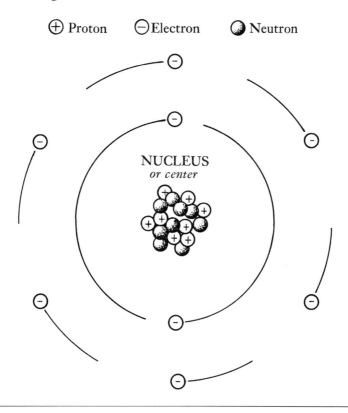

⊕ Proton   ⊖ Electron   ◒ Neutron

NUCLEUS
or center

HOW U-235 SPLITS

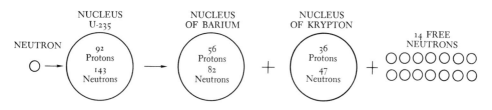

"enriched U-238." Enrichment results from the process of separation which consists of concentrating more U-235 within a given volume of U-238, thus "enriching it" to any desired level.

The mixture of U-235 and U-238 has an added advantage. The nucleus of U-235 will split only if the neutron that hits it is going slow. Since all neutrons travel at enormous speeds, they must be slowed down and this is done by interposing some substance, such as carbon (graphite), called a moderator. On the other hand, U-238 will decay into plutonium only if hit by a *fast* neutron. Therefore, different combinations of U-235, U-238, and moderator give a wide variety of choice in the amount of plutonium produced.

In other words, nuclear reactors can do two things simultaneously: produce heat to make electricity and produce plutonium. The government buys plutonium at a high price from public utilities, thus helping to lower the cost of nuclear-produced electric power.

The uranium fuel is in the form of solid shapes—usually rods which are completely covered with a metal, often aluminum, to contain the radioactivity produced in fission and in changing U-238 to plutonium. These sealed rods are inserted in the moderator. After some time the radioactive by-products are so great that they interfere with fission and have to be removed. The partially spent fuel cells are taken out, stored in special places for a time to get rid of the worst radioactivity, and then sent to a reprocessing center where the radioactive wastes are extracted by means of an acid solution. This acid solution is called a high-level waste—so hot with radioactive decay that it must be kept cooled to prevent boiling. It is ex-

tremely toxic and dangerous; a gallon may run over a thousand curies in radioactivity. This is so lethal that if *three* gallons of it were distributed equally among the world's population the danger point of radiation for the human body would be reached for everyone on earth.*
Yet some *one hundred million* gallons have been produced, and most of it has been buried at Hanford in the state of Washington, in tanks which hold from one-half million to a million gallons each. They are made of concrete lined with stainless steel and require the most elaborate care. They must be kept cooled, the solution must be stirred to prevent radioactive concentrations or "hot spots" that could melt the stainless steel, and the tanks must be replaced every twenty years. This kind of care has to go on for the next thousand years while the radioactivity dissipates!

Despite all care, leaks take place. It is known that up to 1965 at least four leaks have taken place, and that some 60,000 gallons of this lethal brew have seeped into the earth at the Hanford site. The consequences of this will not be known for a long time as the radioactive liquid works its way into the aquifers.

The radioactivity of the brew in the underground tanks at Hanford is such that, if released, it would be the equivalent of exploding all the nuclear warheads in the Amer-

* Estimated by Professor David R. Inglis, a nuclear physicist at the University of Massachusetts, who is also the authority for the estimate of 100 million gallons of hot wastes at 100 curies per gallon, or a total of 10 billion curies. The lowest estimate made is that of the Williamstown Conference which may be considered official because of AEC participation. Their figure is 1.5 billion curies.

The disparity is somewhat irrelevant for two reasons. The lower figure is 100 times more than enough to kill all people on earth, and, secondly, the Williamstown Conference projects *for 1980* a total high-level waste of *44 billion curies!*

ican stockpile. Since their general location is known, in
case of thermonuclear war it is probable that the enemy
would aim a few missiles at the Hanford site. Without
enemy action, nature could take a hand in the form of
earthquakes. The tanks at Hanford were built *without
an adequate geological survey of the area,* and it now
turns out that the area contains many fault lines (a fault
line is a fracture in the earth's crust which gives under
stress). There have been minor earthquakes in the area,
including one in 1918 which may have occurred on the
very site of the Hanford plant. (*Environment,* May, 1970.)

The placing of those tanks is the most startling example
of the reckless disregard of elementary ecological pru-
dence by men in responsible positions under the pressure
of military requirements. It is obvious that had they known
of the danger, the storage tanks would have been located
elsewhere, despite the added problem and expense of transporting the radioactive wastes by tank cars or pipelines.

Whatever excuse the original military planners may
have had of functioning under wartime pressure does not
hold for the Atomic Energy Commission which for 15
years did not even recognize the danger. Belatedly, it is
now doing a great deal of research on disposal of high-
level waste. One approach, already being tested, is to
solidify the waste and bury it in abandoned salt mines. A
pilot plant for waste solidification went into operation in
Hanford in 1966. In a period of three years it had managed
to process 1 per cent of existing waste. Other solutions
are being investigated. One is to inject the waste deep
beneath the earth into porous sandstone layers that would
absorb the liquid. Another is to dig a deep well, say 5,000
feet, and apply pressure to crack the rocks below. Liquid

waste, mixed with a hardening substance, would be pumped in to fill the cracks. Once solid, it would be safe for a long time.

Hitherto, the bulk of high-level waste came from the production of plutonium which has been used almost wholly for weapons. Since both the U.S.S.R. and the U.S.A. have enough weapons for complete mutual destruction, a moratorium on the production of plutonium for weapons would be beneficial to mankind. However, while such an agreement would help, it would not solve the problem, because the reprocessing of fuel elements will go up due to the increase of nuclear power plants. The AEC estimates their output to increase from 10 million kwh in 1970 to 150 million kwh in 1980. These estimates may turn out to be too high, since nuclear power has proved more expensive than expected, and utilities companies have cut back sharply on orders for nuclear plants—from a high of 25 billion watts in 1967 to seven billion in 1969. To stimulate business, the AEC is spending 103 million dollars for fiscal 1972 on research on another kind of reactor which uses plutonium, the so-called "fast breeder" reactor.* This kind of reactor is very hazardous compared to the reactors now in use. Reactors now in use cannot explode; the "fast breeder" reactor can. This makes quite a difference in safety measures to contain the enormous radioactivity in

---

* The more "enriched" the fuel, the more efficient the power plant. Pure U-235 would therefore be the most efficient . . . and the most dangerous. Plutonium works the same way, and if mixed with U-238 will make more plutonium as well as give out energy, so this kind of reactor "breeds" more fuel than it consumes.

The radioactivity in a reactor is fantastic. A 500-megawatt reactor at the time of shutdown for fuel removal may have in its fuel 4 billion curies of radioactivity—a thousand times more than enough to kill every one on earth. In 24 hours this goes down to one-half billion curies, but a hundred years later it will still produce 5 million curies—still a lethal dose.

a nuclear power plant which is almost all kept inside the sealed fuel elements. An explosion can burst those elements open and scatter them over a very wide area.

The scientific community is extremely upset about the "breeder" reactor program, not only because it is so hazardous, but because its funding takes money away from research on fusion technology, which is the great hope of the future (see Chapter 13). In an effort to stir public debate, the Scientists' Institute for Public Information, headed by Dr. Margaret Mead, filed suit against the AEC on May 25, 1971 to force the agency to disclose the impact of the program on the environment as required by the National Environmental Policy Act of 1970. It is again a comment on AEC policies that it should have to be sued to obey the law.

Low-level wastes are much less hazardous than the high-level wastes, but they can constitute a serious problem if their volume increases. As mentioned on page 80, low-level radioactivity from the English nuclear plants is adversely affecting marine life in the Irish Sea. The impact of radioactivity on the ocean is just beginning to be investigated. There is some release of low-level waste from the use of radioactive isotopes in medicine and industry, yet this problem can be controlled by strict governmental standards of disposal. The low-level waste of nuclear plants can also be controlled, but the problem is much greater.

There are three sources of low-level radioactive waste in reactors. One is from the ventilation of the system and is negligible—except for accidents such as the fire at the Rocky Flats plant in 1969. A second source is the water which surrounds the fuel elements and absorbs their heat, thus making the steam which drives the generator's turbines, as shown in the simplified diagram on this page. These fuel elements are sealed in a thin metal bonding to prevent absorption by the water of the radioactive fission products. But there are always leaks in the bonding metal, so that the water in this internal system is radioactive. A third source is the water from a river or lake which is used to cool the condenser coils. This water forms an open-ended cycle returning to the river or lake. In various ways these

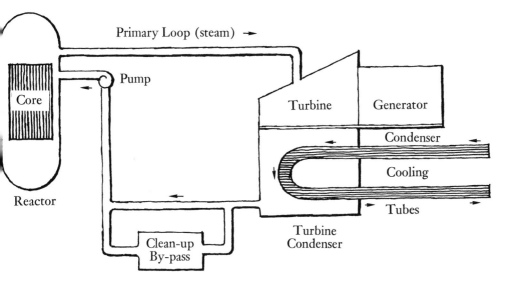

last two sources carry radioactive waste into the lakes or rivers. However, as reactor technology improves, the low-level wastes can, and will be, contained and disposed of in ecologically sound ways.

Aside from radiation pollution, current nuclear plants are responsible for a great deal of thermal pollution, that is, raising the temperature level of the rivers or lakes where they are located to such an extent that fish life is adversely

affected. As the diagram shows, the water which cools
the condenser coils is returned to the river or lake at a
high temperature. For every kilowatt of electricity produced, a nuclear plant produces two kilowatts of heat.
As more nuclear plants go into operation, their heat becomes a significant source of water pollution. The following example and estimates will give some idea of the
magnitude of the problem. There are 16 nuclear power
plants now operating or on order for location on the
shores of the Great Lakes, all scheduled to be operative
by 1975. They will give out some 25 million kilowatts
of heat. This is the equivalent of heating the Mississippi
River by 15 degrees every day and diverting it into Lake
Michigan. What the effect would be on the plants and
animals and ultimately on the people living on its shores
is a matter of conjecture—but it wouldn't be very good.
Fossil-fueled power plants (coal and oil) are also heavy
thermal polluters, although about 50 per cent less so than
nuclear plants because they use less water. In the coming
decade, engineers expect to have more efficient nuclear
plants with their thermal pollution on a level with conventional plants. On the other hand, the conventional
plants have a much higher level of air pollution. Both
types of plants will greatly increase over the coming
decades. The 1970 capacity of all power plants was 300
million kilowatts; this will rise to 550 million kilowatts
by 1980, and to 1,000 million kilowatts by 1990. With
such increases, thermal pollution is coming to the fore as
a major problem. In lakes, for example, much of the
microscopic plant and animal life is killed by the temperature change as it goes through the cooling systems. As
the temperature of water rises, oxygen dissolved in water

is driven off and at the same time bacterial life is enhanced, further depleting oxygen. The result is eutrophication. In the sober words of one American authority, "The United States is approaching the environmental limits that its water resources will tolerate as coolants for waste heat." (Ralph E. Lapp, *New Republic*, February 6, 1971.)

Thermal pollution can be reduced to safe levels by the use of various types of cooling towers and/or evaporating ponds. Cooling towers are tall structures, some 400 feet high, where the hot water is circulated and cooled.* They add to the capital costs of the plant, but not prohibitively —about 8 to 15 per cent. Evaporating ponds are cheaper if there is sufficient land available near the power plants.

The government has moved vigorously to ameliorate thermal pollution. It has set a standard for inland water of not more than a 5-degree increase over monthly averages, allowing maximums of 86 to 96 degrees for short periods of time. The ceiling is 55 degrees on trout and salmon streams. On Lake Michigan, where the new nuclear plants would greatly pollute the waters, the federal government on May 7, 1970, forbade the discharge of water heated more than *one degree* above the existing temperature at the point of discharge. The strict rules are being contested by the courts by some companies such as the Florida Power Company, which has been discharging hot water into Biscayne Bay, but such suits are essentially delaying actions. The trend is clear: thermal pollution can, and will be, controlled.

If both thermal pollution and air pollution can be re-

* General Electric has developed a more efficient cooling tower that would look like a stadium and shoot plumes of hot air thousands of feet upward, literally shooting "holes" in any inversion layer. (*New York Times*, April 7, 1971.)

duced, it would seem that conventional power plants are
preferable to nuclear plants. Radioactivity is very dangerous and its impact on the environment yet to be fully
assessed. We now have barely 20 small nuclear plants
in operation; the AEC projects 600 large ones in the next
30 years. The question of nuclear safety, already worrisome, will be of increasing concern and cannot be left
in the hands of the agency whose main interest is the
promotion of nuclear power, "to spread the gospel of the
peaceful atom," in the words of Howard Brown, a top
executive of the agency.

The AEC has consistently underestimated the dangers of
radioactivity, and it still has a very cavalier attitude. A
speech of AEC's chairman, Dr. Glenn T. Seaborg, was
excerpted in the *New York Times* of December 28, 1970,
in which he dismissed the dangers of radioactivity: "We
are receiving, and will continue to receive, only a small
fraction of a level of radiation that, based on the best
evidence available, has never been proved harmful to us."
Since the "best available evidence" is not very good or
extensive, the statement is misleading.

Commissioner Seaborg was taken to task by Senator
Mike Gravel of Alaska in an answer printed in the *New
York Times* on January 11, 1971. He points out that "it
is intellectually devious for nuclear enthusiasts to limit
their remarks to today's radiation exposure . . . when two
AEC Commissioners testified last year that . . . they have
no idea what the big plants will actually put out."

A sober discussion of the problem of nuclear safety was
presented in *The New Republic* of January 23, 1971, by
Dr. Ralph E. Lapp. Dr. Lapp is one of the foremost
American authorities on nuclear matters. His article, "How

Safe Are Nuclear Power Plants?" discusses the inadequacy of existing data and the AEC's public relations approach to scientific questions. His criticism is the more powerful for its restraint: "When AEC Chairman Seaborg says: 'We know that the benefits we gain will far outnumber the risks of the potential hazards,' then I submit the 'we' is not all of 'us.' I do not make the charge that the AEC is imposing an unsafe system of nuclear power in the nation; I submit that the public record is not visible to assure public confidence in the AEC's assurance."

Public criticism of the AEC finally bore fruit in the summer of 1971 with the resignation of Chairman Seaborg. The new chairman, Professor James R. Schlesinger, took office on August 16 and in his first press conference said that the agency would henceforth be "wholly forthright" on its dealings with the public on environmental problems and would release "large amounts of technical information." After this conference an AEC official told the *New York Times* that the new chairman was "absolutely determined to open up the commission." Yet the bureaucracy of the AEC is so allergic to public debate that the new chairman's policies may be frustrated.

Hence, a reorganization of AEC seems advisable. Dr. Lapp, Senator Gravel and many other critics argue that the AEC should be divested of its regulatory functions, which should then be vested in the Environmental Protection Agency. They also agree that there are other alternatives to nuclear power plants (to be discussed in chapter 13) and it would therefore seem prudent to decelerate the massive expansion projected by the AEC.

● 7 ●

WHERE EVEN THE BIRDS COUGH

Wherein a joke is no longer a joke. Smog—the killer of people and plants. Some major polluters of the air, and a good look at the worst offender, the automobile. Some reflections on cars, power plants, and supersonic transports. Some ideas on solutions to air pollution.

Many, many years ago, people used to joke that Los Angeles was a place "where even the birds coughed." It seemed rather funny in those days. But in October of 1948 a sudden illness struck a small Pennsylvania town, laying low one half of its total population of 13,000 people, and killing 20 of them. It seemed like an epidemic, but it wasn't. It was the result of poisoned air. Who was to blame? Who but the townspeople themselves, whose factories and cars had been creating the smog that killed them. When the killing smog settled on London for a three-month period in the winter of 1952–1953, 12,000 more deaths than normal were registered and attributed to the smog. The joke wasn't funny any longer.

The overwhelming bulk of air pollution comes from combustion: the burning of wood and trash in dumps, of coal and oil in furnaces, of gasoline and oil in engines. When primitive man burned wood the pollution was dissipated into the air. Then came coal, and as early as 271 B.C. a pupil of Aristotle complained of the objectionable smell of coal. Yet pollution was not a problem except in very large cities. In ancient Rome, for example, aristocrats bewailed the effect of soot on their white togas, and

British royalty in the Middle Ages complained about coal smoke in London. In 1257, Henry III's wife, Eleanor of Provence, moved away from the court to a rural castle because of the "unendurable" smoke in Nottingham, and that led 13 years later to a parliamentary act prohibiting the burning of soft coal. Despite the law, urban air got worse, and in 1307 Edward I appointed a commission to enforce the law in London. He even went so far as to put one artificer to death for burning coal instead of oak. The great Queen Elizabeth objected strongly to the smoke, and in 1690 William III left London to escape it.

With the coming of the steam engine and manufacturing, the use of coal spurted upward, century by century. Then came petroleum, whose use has expanded enormously decade by decade. As industrialization increased, so did urbanization and the concentration of pollution over cities. By the middle of the twentieth century, man awoke to the fact that he was being poisoned by the air he breathed—by smog.

The word "smog" was first used by Dr. Harold Antoine Des Voeux at a public health conference in London in 1905, to describe the combination of smoke and fog for which London was notorious. But the word is a misnomer, for the condition it describes is found in cities where there is no fog. Actually it is a chemical pollution which has become concentrated because of atmospheric inversion, a weather condition in which a layer of cold air prevents warm air from rising. In effect, this traps the chemical pollution.

Chemical pollution is chiefly the result of combustion of oil and coal, which releases compounds such as sulphur dioxide or the oxides of nitrogen. Under the influence of sunlight, these compounds combine with hydrocarbons in the air to form ozone and other organic compounds called "photochemical" substances. These substances are particularly harmful to plants. Leafy crops, such as lettuce and spinach, can no longer be grown in areas of southern California. What the ozone can and will do is shown by the following example.

On April 2, 1970, the *New York Times* carried a story of an ecological disaster under these headlines: "ONE THOUSAND ACRES OF SMOG-AFFLICTED PINES TO BE CUT: Giant Ponderosa Poisoned in National Forest by Air of Nearby Los Angeles." The story describes the cutting of those forest giants, two feet in diameter and 75 feet high, which take a hundred years to grow to maturity. They were poisoned in a few years by the smog from the valley. The ozone attacks the tips of the long green needles, which turn brown. This slows down photosynthesis, and the tree's production of pitch —its life blood—also slows down, making the tree an easy prey for bark beetles and other pests which gradually kill it. Of 160,000 acres of pines in the National Forest, two-thirds are already damaged, 46,000 acres severely so. The story goes on to say that, in addition to trees, crops on 11,000 square miles of farmland have been affected for a total damage of a hundred million dollars a year in California alone. The story ends by saying that smog damage to crops has been noted in 20 states and that in New Jersey every single county has reported damage involving 36 commercial crops.

Besides the "photochemicals," many kinds of corrosive acids are produced after combustion. For example, sulphur dioxide combines with oxygen and moisture to produce sulphuric acid which attacks stone, paints, and metals. The facade of Strasbourg Cathedral in France has been so damaged that new stonework is now replacing the old. The marble facade of New York's City Hall had to be replaced a few years ago at a cost of four million dollars. Cleopatra's Needle, the obelisk behind the Metropolitan Museum, has suffered more in the time it was there— less than a hundred years—than in the 3,000 years it was in Egypt. If pollution can peel off layers of stone, what does it do to lung tissue?

That air pollution is dangerous to health in human beings is accepted by all authorities, although a great deal of research is necessary to determine the extent and the mechanisms of the damage done. In Los Angeles, for example, a group of mice breathing the regular city air developed one and a half times as many lung cancers as a control group of mice which was given purified air. Men aren't mice, of course, but the findings are suggestive when one considers the fact that the incidence of lung cancer increases with the size of a city. Generally, the bigger the city the greater the air pollution, and while scientists cannot prove it—as yet—they believe that there is a relationship between air pollution and lung cancer as well as respiratory diseases. In New York City, for example, there was an 80 per cent rise in deaths due to respiratory diseases between 1930 and 1960, whereas the population had increased 10 per cent. New York City air is so polluted that the city prohibits its use in compressed-air tanks for underwater work except by special permit.

In addition to the more obvious danger in respiratory diseases, scientists believe that air pollution contributes to irritability, melancholia, and other nervous and psycho-

logical diseases. The public has accepted the idea that
cigarettes are dangerous to health and many people have
given up smoking. Yet if you live in Manhattan, you
breathe in, every day, toxic material equivalent to 38
cigarettes a day—and you can't give up breathing!

As in other instances of pollution, the sources of air
pollution are steadily on the increase. The amount of
fuel combustion in the world is doubling every 20 years;
in the United States it is doubling every 10 years! Today
in the United States 400 billion pounds of pollutants go
up into the air every year—count them, 400,000,000,000
pounds! This is the estimate of *Fortune* magazine; *Medical
Journal* estimates emissions at 350 billion pounds. Both
agree on an estimate for the automobile of 160–170 billion
pounds a year, or almost half of the total. Power plants
contribute 13 per cent of the total, but within that percentage are 50 per cent of all sulphur dioxide pollution
and 27 per cent of all the nitrogen oxides.

Every minute, nearly one million pounds of emissions
pollute the American air, including 130,000 pounds of
sulphur dioxide, 240,000 pounds of carbon monoxide,
2,500 pounds of nitrogen oxides—every minute, hour after
hour, day after day—every week, every month, every
year. The cities produce the bulk of pollution, and, of
course, get it right back. New York City, rated highest
in pollution, gets a downfall of dust and soot amounting
to 1.5 million pounds a day! It is impossible to keep a
blouse or a shirt clean for more than a couple of hours
in the center of Manhattan. This fallout is only a small
part of the pollution which remains in the air to be
breathed, such as the 3 million pounds of sulphur dioxide
which are emitted daily over the city. In 1969, a city

ordinance required the use of fuel with lower sulphur content. It is hoped this will cut sulphur pollution in half —to *only* 1.5 million pounds a day.

Air pollution is costly not only in medical bills, but in repair bills for materials. Pollution corrodes metals: steel 30 times as fast as in clean air, nickel 25 times as fast, brass 8 times, copper and aluminum 5 times. Paints, wood, stone, bricks, everything is affected. A barber in Pontiac, Michigan, who had worked in many other states and had kept a careful check on the wear and tear of his instruments, found out that they wore out much faster in the smog of Pontiac.

The costs of air pollution in health, medical bills, and repairs have been estimated at 18 billion dollars a year by the government economists at the National Air Control Administration. This is a significant figure, since it has been estimated that air pollution can be controlled at a yearly cost of one or two billion dollars—*exclusive of the automobile* which presents a particularly tough problem. Air pollution can be controlled at a much lower cost than water pollution, and great progress can be made in a short period of time as shown by the experience of Pittsburgh.

Thirty years ago, Pittsburgh was known as the Smoky City—number one in America for air pollution, which came from steel, coal, and railroads, as well as homes, schools, churches, hotels, and so on. There was such a haze over the city that it was a rare occurrence when the sun could break through.

In 1942 a civic group, the United Smoke Council, was organized, representing 71 community organizations. In the following year they got the first antismoke ordinance passed. The group agitated and educated, and in 1945 became part of a large antipollution organization. In the same year David Lawrence ran for mayor and won, chiefly on the issue of smoke control. He showed how costly air pollution was to the city. Pittsburgh led the country in deaths from pneumonia and other respiratory diseases. Forty industrial firms had decided to leave the area because of air and water pollution. A department store estimated that one day of smog had cost it $25,000 in cleaning bills and markdowns for soiled merchandise.

Slowly controls spread. In 1947 the Pennsylvania Railroad agreed to convert from coal-burning locomotives to diesels. Riverboats converted from steam to diesels. Factories, apartment houses, hotels, all were forced to burn smokeless fuels or install smoke-consuming devices. By 1952 industry and communities in Allegheny County were engaged in a 200 million dollar continuing program of air pollution control. In that year it was estimated that Pittsburgh saved 26 million dollars in cleaning bills alone—5 million dollars of that in laundry bills. Visibility was up 77 per cent since 1945.

Since 1952 there has been an almost steady drop in dustfall, until today it is half that of New York City. In the past decade 132 open-hearth furnaces have limited smoke and fly ash and are working to screen dust from gas. Pittsburgh isn't perfect by any means, but it has made enormous progress and has created a momentum which will probably enable it to show the way to other large cities.

Technically and economically, air pollution can be overcome—whether it comes from power plants, from industry, from municipal and private incinerators, or from institutional and domestic heating. The technology for smoke control has been available for many years, but recently systems have been developed which eliminate both the soot particles and the noxious oxides of sulphur and nitrogen, as well as certain radioactive emissions. Most notable of recent developments of this type has been the Hydro Precipitol which the Harvard school of Public Hygiene has praised for its radically novel approach. The Hydro Precipitol is the invention of the noted aircraft designer, Major Alexander de Seversky, who has long been concerned with pollution, and who turned his talents to finding solutions. Size for size, the system is claimed to be 100 times more efficient than conventional electrostatic precipitators and a 1-million-dollar installation was being set up in January, 1971, at the lead smelting and refinery plant of the General Battery Corporation in Laureldale Township, Pennsylvania. Also in production is a unit for a power plant of Southern California Edison Company.

Another innovation has been the Ironton preheaters for coke-producing furnaces, which have been one of the more difficult problems in pollution controls. The new system cuts smoke and gases by as much as 70 per cent and at the same time increases production by 50 per cent. (*New York Times,* January 13, 1971.)

In the words of Federal Administrator William D. Ruckelshaus (Environmental Protection Agency), "There is no excuse for delay in applying proven means of control." Speaking to an audience of businessmen at the National Press Club in Washington, D.C., on January 12, 1971, the administrator pointed out that power plant emissions could be reduced from 5.6 million tons a year to less than 80,000 tons; iron and steel emissions could be cut down from 1.5 million tons to 30,000 tons; the rest of industry, apartment houses, and commercial buildings could cut down from 17.5 million tons of emissions to 700,000 tons. In all, smoke pollution could be cut down a whopping 96 per cent! "Not to clean up smoke," he stated, "is inexcusable."

If citizens become aroused, smoke pollution can be eliminated in a few years. But the major problem of air pollution, created by the automobile, remains and this is a difficult problem indeed.

The great difficulty with the automobile, of course, is that it is so deeply embedded in the social and economic life of the United States that it cannot be eliminated. Nor can it be easily changed: the economic pressures alone are tremendous.

The automobile and its ancillary industries (tires, gasoline, highways, auto insurance, service stations, repair shops, credit organizations, parking lots, trucking) constitute a veritable kingdom, a state within a state, that staggers the imagination. The income and expenditures of General Motors alone are greater than the national budgets of any nation on earth except the U.S.A. and the U.S.S.R.! A rough estimate of the value of goods and services related to the automobile for 1970 is around 150 billion dollars or about a fifth of our total national income. Again, very roughly, this means that a fifth of the population depends on the automobile. It also means, at these rough levels, that one fifth of all pollution and ecological problems relate to the automobile—in some cases more. For air pollution, the figure is 50 per cent, but that is only the *direct* pollution of the 90 million cars in the United States today. Their manufacturing creates pollution, and so does the production of the materials in the car—the steel, glass, plastics, textiles, paint, wiring, batteries, etc. The oil refineries that provide the gas and the rubber industry that makes the tires create pollution. All these industries need power, and power is a major source of pollution.

But this is only air pollution. There is water pollution and solid-waste pollution and all types of ecological damage, whether it is highways driven through wilderness forests or big sections of fertile farmland carpeted in concrete. The damage of highways which go through cities is only now beginning to be appreciated as our cities choke on cars.

Air pollution from cars can certainly be cut down, but how much is an open question. The auto companies assert that they can clean up the internal combustion engine, but they fought bitterly the Muskie Bill which gives them six years to make engines 90 per cent clean. Few people believe that this can be accomplished by present engines, and new types must be developed, such as electric engines, gas turbines, etc. One suggestion is development of a "hybrid" automobile, a mix of electric and gasoline sources of power. The car would be driven by electric batteries which would be recharged in open country by a small combustion engine going at a constant speed and equipped with afterburners to emit only a fraction of the pollutants now given out by conventional engines.

It is obvious that a major approach in controlling air pollution would be to cut down on the number of automobiles in the United States. We now have 90 million cars (about one and a half per family) compared to 60 million in 1955. In other words, cars have increased by 50 per cent in 15 years. If we keep up the same rate we will have 135 million cars by 1985—which, from any point of view, is sheer insanity. It need not happen if attention is focused on increasing and improving mass transit on one hand while cutting down on auto production on the other.

A system of steep federal taxation of automobiles and gasoline (in England a gallon of gas is so taxed that it costs double the price in the United States) would cut down on the number and use of automobiles, and those taxes would be returned to cities and communities for the express purpose of improving mass transit systems. Highway construction should be tapered off and emphasis placed on good maintenance, safe access roads, and enhancing the surroundings. Tolls on bridges, tunnels, and superhighways should be steeply increased, the money again going to mass transit. It should not be technically difficult, nor socially onerous, to reduce the number of cars to some 60 million within five years, roughly one car per family. The combination of taxation and cheap, rapid, comfortable mass transit should do the trick, without major dislocations.

Of course, mass transit means power, a great deal of it, so that the reduction of pollution from automobiles is offset by the increase in pollution by power plants. Nevertheless, the saving in pollution is very great. An economist, Bruce C. Netschert, has done the pencil and paper work, using 1968 figures, to assess how much more electric power would be needed if all automobile traffic were to be carried in electric automobiles. At the same time, he figured out the ratio of electric use between auto and rail transport, and finally estimated the amount of electricity needed if the railroads were electrified. He eliminated from his calculations the thousands of miles of branch and spur roads and used only the main lines (some 55 per cent of the total) which carry about 90 per cent of all rail traffic. His work was published in *The Bulletin of Atomic Scientists* for May, 1970, and here are the comparisons:

AIR POLLUTION (*in millions of tons per year*)

Before Electrification
Due to automobiles	85
Due to railroads (*diesels*)	2
	87

After Electrification
Electric automobiles require an additional 616 billion kw	11
Electrified rails require an additional 32 billion kw	.5
	11.5

Net saving in pollution   75.5

Therefore, if all autos were electric the net saving in pollution would be 75 million tons a year, or about half of all present combustion pollution of 170 million tons a year. These figures are of course impossible to achieve, as there is at present no electric car acceptable to the American people. Existing battery-driven cars have a speed of 35 miles per hour and a range of 50 miles before recharging. A practical car should do 50 miles per hour with a range of 150 miles. Japan has allocated 14 million dollars to developing such a car within five years.

However a lot can be done, and we can outline a plan to cut auto pollution by two-thirds within 10 years based on a crash program for a high-grade mass transit system, a practical electric auto, and a 40 per cent reduction in the pollution of the present automobile (the Muskie Bill requires a 90 per cent reduction in six years).

A further assumption for the plan is that electric cars will not be taxed at all, while regular cars (new or used) pay a 100 per cent federal sales tax—doubling the price to the consumer. There would be a massive drop in cars purchased and a massive shift to electric cars for suburban and urban use. The extent of the changes is arguable, but a 20 per cent drop in each category is not unrealistic. Since there are 90 million cars at present in use (1971), we are assuming the elimination of 18 million cars and a transfer of 15 million to electric types, leaving 57 million standard-type cars or one such car per family. Since this was the average as recently as 1955, the assumption would not create too big a shock to our customs. Here are the results in terms of pollution:

	AUTO POLLUTION DROP IN (*millions tons yearly*)	INCREASE IN POWER PLANTS POLLUTION
20% fewer autos	17	0
20% shift to electric	14	2
40% reduction of pollution in remaining autos	21	
total	52	2

On this scheme, within ten years, 50 million tons of annual pollution would be eliminated, roughly two thirds of present auto pollution, and 30 per cent of all combustion pollution. With industry and power plants doing their share, it is not too unrealistic to project a reduction of combustion pollution by half in 10 years—if people push for it. During this time a completely new internal combustion engine or some suitable alternatives could be developed, eliminating another 35 million tons of pollution a year. By that time, 10 years hence, the worst of the old factories and power plants would be retired, new construction would have to meet ever more stringent requirements, and at the end of another five years perhaps another 10 to 20 million tons could be lopped off, leaving perhaps 30 million tons (or one sixth of present pollutions) for future generations to work on.

Such a fifteen-year phased program, or variations thereof, is in no way utopian: it is economically and technically feasible, and only awaits the political muscle of a concerned electorate.

If nothing is done, or too little is done, air pollution in the United States within twenty years will be enormous. Power capacity will triple to one billion kilowatts, of which half will be coal-fueled, roughly doubling pollution from 25 million tons to 50 million tons. Autos (even with pollution controls) will add another 50 million tons; and if industry, heating, incineration, etc., keeps pace, we can look forward to some 400 million tons of air pollution within 20 years. The cities will be unbearable.

Of course, something will be done. But how much? Compared to water pollution, the cost of cleaning air pollution is much lower, perhaps a tenth—provided something is done about autos and power plants. Among the more important steps would be one to persuade the auto industry to build cars that last longer. The present life of a car is about three years, or 75,000 miles, without repairs, and an additional four years and 75,000 miles with considerable repairs. Without question, cars can be built with double and triple that life span at only a small extra cost—perhaps 10 or 20 per cent additional cost. The lower profit margins could easily be subsidized by the government from the sales tax previously suggested. Of course, longer lasting cars would raise serious adjustment problems for companies whose sales thrive on obsolescence. The higher cost would cut down the number of cars bought, and the longer life could cut down the disposal problem of old cars, which is now becoming acute. As important would be the saving on raw materials (not a negligible benefit), and the saving on pollution of the intermediate industries—steel, glass, power, etc.

What is true for cars is true of all metal products, particularly aluminum, which uses up 10 per cent of all power production. Beer cans and other containers thus add to air pollution as well as to solid-waste disposal problems, which we will discuss in Chapter 9. Nitrogen fertilizers use a great deal of electric power and so do paper pulp mills. Cutbacks on those industries, by using more thrifty methods, would have multisided benefits.

In general, the planned obsolescence now standard in American industry is a major offender in all pollution problems. If things are made to last, there are fewer problems all the way down the line, from less production to less waste disposal. The case of the automobile is especially important, because of its role in the economy. The suggestions made will be resisted as undermining the American standard of living. But if by the American standard of living we include the quality of life, then it will not be undermined; it will be strengthened.

• 8 •

THE PROS AND CONS OF PESTICIDES

Wherein we speak of chemicals that pollute soil, air, and water in every part of the world. Their evil and their usefulness; why men of good will argue on both sides of the question.

Pesticides and herbicides are chemicals usually known as insect-killers and weed-killers (*cide* comes from the Latin *caedere*, to cut down or to kill; hence homicide, suicide, genocide, terracide, patricide, etc.). The latinized words are more accurate since pests include animals other than insects (larvae, rodents, birds), and herbicides kill any green plant as well as weeds.

A pest is whatever competes with man for his food and fiber supply (for example, by destroying cotton in the field or in storage) or interferes with his health or comfort (malarial mosquitoes, bedbugs, etc.). Sometimes a pest in one place is a boon in another: a bee in the house is a nuisance; it is an asset in an orchard. Similarly, a weed is simply a plant in the wrong place: the dandelion in your lawn is a nuisance; in a meadow it is a source of food to cows, of salads and wine to humans. Of course, there are many pests that are pests everywhere: it takes a real lover of nature to say anything good about the anopheles mosquito.

The usefulness of pesticides is beyond question. Fairly early in history we find man burning sulphur to fumigate homes and storehouses, using arsenic to poison rats and weeds. By the twentieth century man had evolved many types of insecticides derived from minerals (compounds of arsenic, copper, lead, zinc, etc.) and from plants—pyrethrum from chrysanthemums, nicotine from tobacco, rotenone from leguminous plants.

In recent times, particularly since World War II, synthetic pesticides have been developed. These are chemical compounds not found in nature, but synthesized by man (from natural elements, of course) in his laboratories, chiefly as a result of tests for chemical warfare where insects are used to test for the toxicity of new compounds.

These synthetic pesticides are much more powerful than the old ones. This is all to the good in killing pests, but many have such enormous biological potency that even infinitesimal amounts are harmful to animal life from fish to man. Let us take DDT* as an example, for it has been the most widely used pesticide and is representative of the wider problem: the conflict between the beneficial and harmful aspects of the new synthetic pesticides.

DDT is shorthand for a long, complicated chemical name: dichloro-diphenyl-trichloro-ethane. It was first synthesized in 1874, but its properties as a pesticide were only discovered in 1939 by Paul Müller of Switzerland. It was immediately used by the United States Army to combat lice, malaria, and other diseases or carriers of diseases. Hundreds of thousand of soldiers, prisoners-of-war and refugees were sprinkled with DDT powder with no ill effects of any kind. Farmers then took up DDT to protect their crops. The pesticide was so obviously successful and so valuable that in 1948 Müller received the Nobel prize.

---

* DDT has close relatives, DDD and DDE, which act in the same way and are often used for a number of reasons. They are called residues of DDT, as DDT breaks down to form them, and all three, DDT, DDD, DDE, are found together in fatty tissues of animals and fish.

In the 25 years following its first use, the achievements of DDT were spectacular. No less than 5 million lives were saved, according to Dr. Edward F. Knipling, Director of Entomology Research at the U.S. Department of Agriculture. In addition, he estimates that a hundred million illnesses were prevented through DDT's control of malaria, typhus, dysentery, and other diseases. DDT not only saved lives but it saved crops and helped higher production of food. Wherever it controlled malaria, for example, agriculture productivity rose from 15 per cent to 50 per cent. It was largely responsible for doubling the yield of potatoes in the United States. In general, DDT and other pesticides can effect an increase of some 30 per cent to 50 per cent in the yield of most crops. In a world of increasing population and increasing shortage of food, the contribution of pesticides is invaluable and DDT, DDD, DDE seemed ideal. Then doubts began to spread as evidence accumulated that fish, birds, and animals died in places where DDT and its relatives were used. What was happening?

Clear Lake, 90 miles north of San Francisco, was one of the cases that pointed to the answer. It had been treated with DDD in 1949, in a concentration of one part per 70 million parts of water, and the results had been very good: the huge gnat population was drastically reduced. However, by 1954, it had built up again and the DDD treatment was repeated, this time in a concentration of one part to 50 million. The destruction of the gnat was thought complete.

During the following winter, about one hundred western grebes on the lake were found dead. The western grebe, a bird of spectacular appearance, was a frequent visitor to the lake. The bodies were examined for infectious diseases or other causes of death, and none was

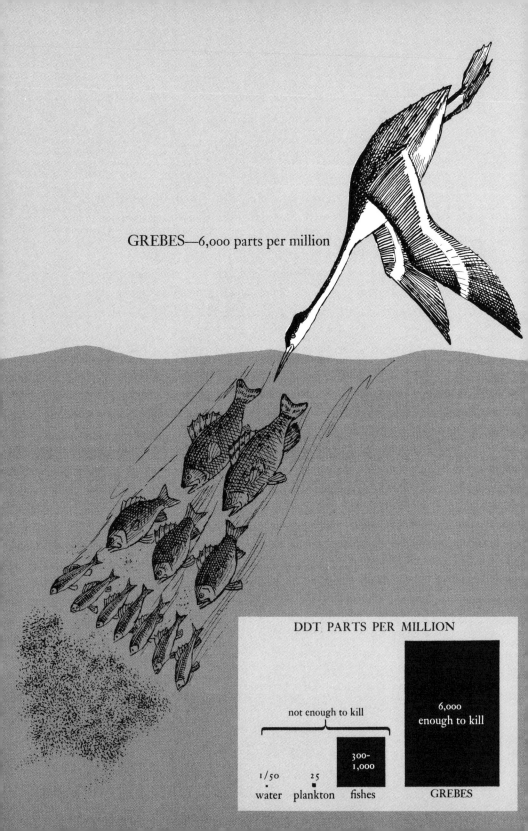

GREBES—6,000 parts per million

DDT PARTS PER MILLION

not enough to kill

6,000
enough to kill

1/50
water

25
plankton

300–1,000
fishes

GREBES

found. In 1957 the gnat population was up again and DDD treatment was again applied. More grebes died. Upon examination, no disease was found, but someone had the bright idea of analyzing the dead birds for DDD: a concentration of 6,000 parts per million was found in the fatty tissues of the dead birds! How could such deadly concentration have arisen from water which contained only 1/50th of a part per million parts of water? The answer was found in the food chain.

The plankton in the lake showed a concentration of 25 parts DDD per million parts of plankton, plankton-eating fish had up to 300 parts of DDD, and fish-eating fish had 10 times that amount (one particularly voracious brown bullhead had an astounding concentration of 25,000 parts per million). It was easy to see how the fish-eating grebes reached a concentration of 6,000 parts per million. Obviously, DDD accumulated in the fatty tissues: each meal of a fish increased the poison in its body, and while it didn't kill the fish, it provided the successive additions that killed the grebes. A small organism might be comparatively safe while a larger organism would die.

This is what happened to the lizards and cats of Borneo described in the introduction. The lizards became sluggish but didn't die; the cats, able to catch a lot of lizards, built up the concentration of poison to a lethal dose; similarly, for the grebes, which were nearly wiped out. Of the 1,000 nesting pairs of grebes in Clear Lake in 1949, only 30 remained in 1960. There were no reports of illness among fishermen who caught and ate fish from Clear Lake; presumably no one had the day-in, day-out diet of poisoned fish that the grebes had. Nevertheless, in 1959 the California health authorities stopped the use of DDD at the lake.

As the years went by, further research confirmed the concentration effects of DDT, DDD, and DDE in the food chains. Not only was the case of Clear Lake not an exception, but higher concentrations were found, particularly in the classic study made in the summer of 1966 by Professor Charles F. Wurster of the State University of New York and George Woodwell of the Brookhaven National Laboratory. They studied intensively a salt marsh on the south shore (the ocean side) of Long Island which had been sprayed with DDT for 10 years as a malaria control measure.

The water in the marsh contained 50 parts of DDT per trillion parts of water (50 parts to 1,000,000,000,000 parts). The zooplankton contained 100 times as much, and the small fish, minnows and silversides, feeding on the plankton, had 10 times as much again, or one part per billion. Larger and larger fish showed higher and higher concentrations, up to one or two parts of DDT per million. The big birds feeding on the large fish, such as the diving ducks and the cormorants, showed concentrations of some 30 parts per million or roughly one million times more than in the water!*

There is no question now that DDT is dangerous. The reason there were no ill effects from DDT among soldiers was that DDT is not readily absorbed through the skin, and people were only dusted with it. The war experience helped to give DDT its reputation for being harmless. But dissolved in oil, the way it is normally used in agriculture, it is highly toxic to man and beast.

Minute quantities in water and food accumulate in the human body, and in some areas of the United States the level of DDT in mothers' milk now exceeds the standards of

* Concentration of DDD in the grebes of Clear Lake had been "only" 10,000 times greater than in the water.

safety prescribed by public health authorities. The successors of DDT in the same family of chlorinated hydrocarbons, such as chlordane, heptachlor, dieldrin, aldrin, and endrin, are more poisonous than DDT and all are stored in the fatty tissues of organisms. The other major family of pesticides, the organic phosphates, are even more poisonous than the hydrocarbons but, fortunately, decompose rather rapidly and therefore their residues on crops and soil are short-lived. They can still do significant damage, however. A 1969 study in Tulare County, California, in which 58 children of farm laborers were tested for malnutrition, turned up signs of low-grade poisoning by phosphate in 27 children—practically half. Parathion, one of the phosphates, is so powerful that spray operators have to be carefully protected and periodically examined. California reports some 200 deaths a year from accidental parathion poisoning; Japan, over 300.

Malathion, another organic phosphate, is widely used as a household insecticide, because it was long presumed to be "safe." It has been discovered, however, that it is safe only because the liver renders it relatively harmless as it passes through the body. The detoxification is done by one of the enzymes of the liver, and if this enzyme is destroyed, the person involved gets the full force of malathion. There are chemicals which do affect this enzyme, and chlorinated hydrocarbons which are often mixed with phosphates are among them. Therefore it is possible to get a mixture in which each chemical is only mildly toxic, but together are deadly. Moreover, such combinations seem to multiply the toxicity of each agent by 40 or 50 times. Finally, the phosphates can interact in the body which has been taking chemicals from other sources: thus, malathion increases the effects of barbiturates, parathion increases the toxicity of muscle relaxants, and so on.

This problem of interaction of chemicals is a serious one; how serious no one knows, as we are just beginning to get a glimmer of its scope. Harmless or comparatively harmless chemicals may combine spontaneously to create harmful compounds. For example, in 1943, the Rocky Mountain Arsenal of the Army Chemical Corps, located near Denver, began to produce war chemicals. The arsenal discharged its waste chemicals (chlorides, chlorates, arsenic, fluorides, etc.) into holding ponds. In 1951 the Army facilities were leased to an insecticide manufacturer who continued to use these holding ponds. Meanwhile, even before insecticides were manufactured, farmers several miles away began to report illness in livestock and extensive crop damage. Since the irrigation water for these farms came from shallow wells, government agencies tested the wells in 1959 and found many of the chemicals mentioned above. They guessed that there had been seepage from the holding ponds into the groundwater below, which in eight years or so had traveled the few miles to the neighboring farms.

The most significant fact about this episode however, was that the testing showed the presence of the herbicide *2, 4-D* both in the wells and in the ponds, but this herbicide *had never been manufactured* either by the Army or by the chemical company that had leased the facilities. After careful study, chemists concluded that the insecticide *had been formed spontaneously* by the action of the sun, air, and water on the chemicals discharged into the holding ponds. Scientists began to realize what vast, unknown dangers might be lurking in the interaction of waste chemicals, in themselves harmless, as they unite in the presence of catalyzing air and sunlight.

This episode also underlines the lack of sufficient knowledge on possible side effects of the many synthetic chemical compounds which do not occur in nature. We don't know what interactions may be taking place in our own bodies as we take chemicals unknowingly from air, water, and food, as well as knowingly in the shape of various medications. Most important, we have very little knowledge as to what happens in the soil to various microorganisms, bacteria, earthworms, and so on, without which there can be no agriculture. Experiments have shown that many pesticides interfere with the nitrogen-fixing bacteria. Earthworms seem to survive, but are sufficiently poisoned so that birds die from eating them.

Poisoning of the soil can be a very serious problem, in part because pesticides persist in the soil for considerable time (from 2 to 12 years), and partly because the pesticides have a tendency to accumulate and to concentrate. Tiny wind currents or terrain features affecting water run-offs can concentrate pesticides in an erratic manner. Some, like DDT, have the property of co-distillation with water, that is, as water evaporates and there is DDT in the vicinity, the DDT will evaporate with the water. Thus, long after the DDT has settled on the ground it may, and will, evaporate and move to settle again elsewhere. The result is that an area sprayed with a "safe" amount of DDT can show horrendous concentrations. For example, orchards that have been sprayed several times a year for a few years with the "safe" amount of one pound per acre should have shown a concentration of, at most, 10 pounds or so per acre. But there were cases where concentrations of 50 pounds per acre were common and in one case there was a concentration of 1,300 pounds per acre over several acres!

The pervasiveness and extent of pollution by pesticides and herbicides cannot be exaggerated, since their use has spread over the whole world. United States production went up from around 100 million pounds of pesticides in 1946 to over 1 billion pounds in 1966, a tenfold increase in 20 years. One tenth of this production was DDT (120 million pounds in 1967) of which two thirds was exported.

As the disadvantages of pesticides became apparent, a controversy began and spread on the pros and cons of using them. Against, were conservationists and ecologists; pro, were the pros, or professionals, in agriculture and public health, who saw the benefits of these chemicals and had come to depend on them. They were supported by the chemical companies who made these pesticides and whose economic stake was not small—somewhere between a quarter- to a half-million dollars. The chemical companies were particularly active against the book, *Silent Spring*, by Rachel Carson, which was first published in 1963 and caused a tremendous sensation, selling into the millions of copies. *Silent Spring* brought together the best available information on pesticides and their effects, and while many of the possible hazards were conjectured, there were enough solid facts to have a devastating impact. The book was a major contribution to the public concern with ecology which flowered in the late 1960's.

As the controversy grew hotter, the fruits of ongoing research further bolstered Miss Carson's contentions. Not only did the risks and dangers of the pesticides become apparent, but their usefulness came under attack. It became clear that pesticides did not offer simple solutions: the more the pesticides were used, the more they *had* to be used in the following years. If a farmer sprayed twice in one year, he would have to spray three or four times the next year, five or six times the year after that. Cotton farmers who sprayed three times a year had to spray nine times the following year. Very often repeated sprayings came to be useless; a more potent pesticide had to be used. It was like drug addiction; once you're hooked, you need more and more, stronger and stronger doses.

What had happened? The insects developed resistance to specific pesticides, and the survivors multiplied with impunity. One of the characteristics of insects is the rapidity and extent of their procreation. Thomas Huxley once estimated that a single female aphid (a tiny plant louse that reproduces without mating) could in one year produce so many aphids that their weight would be equal to the weight of all the Chinese then living.

An insect population could be almost all destroyed, but the few survivors could then multiply so rapidly that they would soon be back to the original level. When DDT was used, the entire insect population seemed to be exterminated (and extermination became a favorite word of pest-control experts), but, in fact, a few always survived, either because they had a built-in tolerance to DDT or because they developed one. In the course of several sprayings, the population grew more and more resistant; the amount of spray had to be increased, but the resistance also increased until DDT was no longer useful. At this point, another more powerful and different pesticide would have to be used, such as dieldrin. This has happened in malaria control. The common housefly has developed resistance to practically everything. In the short run, pesticides were killers; in the long run they became harmless to the given pest.

But worse was to follow. The use of pesticides often opened careers as pests to hitherto harmless insects. The reason is complex, but fascinating. All organisms have other organisms that feed on them, and insects are no exception. There are over 700,000 identified species of insects, and in numbers they probably make up as much as 80 per cent of all earth's creatures. If they weren't busy eating each other, they could take over the earth in a matter of months and there wouldn't be enough men and chemicals to kill them all.

It is a common characteristic of all eating sequences or food chains that the weight of those who do the eating (the predators) is much less than the weight of those eaten (the victims); the weight of all planktons eaten is more than that of the fish that feed on plankton, and the weight of these fish is more than of the fish that feed on them, and so on. The reason is that at each step of the food chain, energy is lost. This loss can be considerable. For example, it takes seven calories of grass to make one calorie of beef. Applying this to insect pests, the predators of a given species are always much fewer than the pest. When the pest is 90 per cent destroyed by spraying, chances are that *all* the predators have also been killed. Moreover, because of the larger population, the pest has a higher statistical chance of having some of its members resistant to the pesticide or escaping exposure. Freed from its predators, survivors can multiply much more rapidly than before.

But this is not all. Usually predators keep many insects in check. For example, the predators that ate the bollweevil also ate the bollworm. The bollweevil was a major problem in the cotton fields, the bollworm was a minor one.

DDT was used against the bollweevil. It didn't kill all the bollweevils (the number of sprayings has to be increased each year), but it killed so many of the predators that the bollworm population exploded and has now become a major pest in the cottonfields. The same thing happened to mites, which were harmless until DDT killed the insects prey-ing on them, now the mites are devouring crops—a full-fledged pest.

As the risks and disadvantages of DDT and other pesti-cides began to outweigh their advantage, the ecologists gained the upper hand over the pros. In Sweden and Switzerland dieldrin was banned completely. In the United States it has been banned in wet areas. In 1969 the United States government banned the use of DDT in residential areas, except for emergency epidemic control, and pre-sumably will phase out that pesticide completely in the next few years. Meanwhile several states have applied stricter controls. In April of 1970 the Soviet Union an-nounced it was limiting all production and use of DDT.

Since there are other alternatives to pesticides (to be discussed in Part III), this might seem like a story with a happy ending. But nothing is simple in human endeavor, and this is doubly true in ecology. Most of the underde-veloped countries have refused to ban DDT. At a meeting of the F.A.O. (Food and Agriculture Organization) of the United Nations in December of 1969, the position of these countries was upheld and the organization went on record that: "F.A.O. will continue to endorse proper, restricted use of organochlorine insecticides."

How can such a decision take place: are these ignorant men? No, they are reputable scientists and administrators, some of world renown.

But if such men do not seem to worry, then is DDT not a serious threat to the environment? Are the arguments we have presented against DDT a tissue of nonsense? No, the arguments are not nonsense and the threat is serious, as F.A.O. agrees and clearly recognizes.

The answer is that these men are caught: caught between the known damage of pesticides to the environment in their countries and on the planet, and the poverty of their countries—not enough food, not enough money, too many diseases. DDT is cheap: 15 cents a pound. In El Salvador a malaria control program using DDT cost five cents per person for the chemical and five cents for the application. If El Salvador were to use other insecticides, the nearest in price would cost 25 cents per person and perhaps 10 cents in application, or three and one-half times the cost of DDT. With the same budget, only a third of the people would be protected, with continued illness and low productivity.

Dieldrin is the main weapon against desert locusts, which in the past have swarmed over an area of 20 million square miles in forty countries. In Ethiopia alone, in 1958 the locusts consumed 167,000 tons of grain, a year's supply for a million people. Dieldrin is a long-lasting, dangerous pesticide, banned in many countries. Can Ethiopia do without it?

These scientists and administrators at F.A.O. are saying that in this agonizing choice they are opting for the short-run benefits of using the pesticides and are willing to pay the price of the long-term denigration of the environment.

But, say the advanced nations, pesticides are everyone's problem. They pollute rivers which run to the oceans. The earth is one, and we all live on it. The answer comes back: "Tell us, you western scientists, tell us how many of us must die of malaria or malnutrition—how many Ethiopians, how many Salvadoreans—to help preserve the earth? For whom? For our children and grandchildren? But if we are dead we won't have children and grandchildren—so really you are saying, dear friends, that we should preserve the earth for *your* grandchildren at the cost of *our* lives."

It is a cruel dilemma, and thoughtful people know it can only be solved by the wealthy nations. We have seen what is happening to the oceans, and self-preservation, if nothing else, dictates that we accept the suggestions of the Williamstown work group of eminent scientists:

> "We recommend that as the production and use of DDT are phased out, subsidies be furnished the developing countries to enable them to afford those alternatives that now exist (but are more expensive). Resources must then be made available, both at home and abroad, for an intensive, and cooperative research effort which will seek to develop not only less-persistent pesticides but also, it is hoped, an approach to the control of pests which is, as well, amenable to the stability of ecosystems."

● 9 ●

OUR HIGH STANDARD OF GARBAGE

Wherein we speak of all kinds of garbage. Some figures on the size of the problem. Some approaches to cutting down solid waste. A few hopeful beginnings.

The dictionary defines garbage as any waste parts of food —animal or vegetable—which is thrown away. In general usage the word includes trash, such as glass, paper, plastics, which is collected by the sanitation services. In the large cities, such as New York and Chicago, the sanitation services collect broken-down furniture, refrigerators, air conditioners, etc., and increasingly have to do something about old cars abandoned on the streets. Ecologists call all this solid waste. It costs us some 3 billion dollars a year to get rid of the 2 billion tons of waste produced every year— including 7 million junked cars, 20 million tons of paper, 46 billion cans. Fifteen per cent of total solid waste is household trash and garbage, and it works out to 3.5 pounds a day per family.

Costs are steadily rising as the methods of disposal become more restricted. New York City, for example, used to dispose of its solid waste by three methods: incineration, landfill (inland and coastal), and dumping by barge into the ocean. To stop further pollution of the ocean, dumping has been prohibited, and available places for landfill will be exhausted by 1975. Most cities are looking for dumping places, and a big hole today is worth a lot of money. For example, in England, the Central Electricity Board has paid 18 million dollars to rent the holes of clay quarries (dug by brick-makers over the last 60 years). The Board needs the holes to put the 50 million tons of ashes which its Midland plants will produce in two decades.

Besides dumping and landfill, burning is widely used to get rid of solid waste. For instance, New York City in 1969 disposed of 6.5 million tons of refuse, 1.8 million by incineration, and the rest by landfill. Two huge incinerators are being built which will triple capacity, and this will still leave a million tons of refuse to be disposed of, presumably by recycling. Yet, despite this huge program, New York City is falling steadily behind: refuse is increasing at the rate of 6 per cent a year, which means that in twelve years the amount of refuse will have doubled. At the same time, the incineration plants will contribute to air pollution, although efforts are being made to keep the pollution down. The use of plastics (polystyrene) in packaging has added to the problem by requiring a greater heat level in the incinerating plants—about double what it was twenty years ago.

About one tenth of all solid waste is handled by the cities; the rest dumped near factories and in the countryside, particularly along highways. Every year, 20 million cubic yards of trash are dumped along the highways, making those unsightly junk heaps and dumps which everyone bemoans and few do anything about. As in so many other areas, the automobile is a major offender. In the old days, scrapped automobiles would be stripped for valuable parts and the bodies sold for steel scrap to the steel mills. Modern steel-making uses much less scrap, so the price of scrap has gone down and it is no longer profitable for junkyards to process old automobiles. Increasingly these cars are simply abandoned in the streets and on the highways.

Another serious offender has been the nonreturnable glass bottle and the metal can. In the last few years their use has greatly increased until it is now estimated that we Americans discard each year 48 billion cans and 28 billion bottles. They clutter our dumps, our highways, our beaches, and our countryside.

A third major offender is the use of plastics in packaging, since most plastics are hard to disintegrate. A return to greater simplicity in packaging and the use of materials that readily disintegrate would help to dispose of solid waste, which, it must always be remembered, has already contributed once to pollution problems in the process of being produced.

Occasionally, local conditions may provide ingenious solutions to solid-waste disposal. One of the more interesting is a project now nearing completion in Du Page County, Illinois, where a recreational hill is being made out of garbage—known as Mount Trashmore. Once completed, it will be planted in some sections, while other parts will be used as ski slopes. The man responsible for this idea was John Sheaffer, who was also the moving spirit behind the Muskegon County Project described in Chapter 4. Mr. Sheaffer, in building Mt. Trashmore (see diagram on opposite page), refined and adapted to local conditions a successful experiment made in postwar Berlin, where the rubble of bombed buildings was used to build up a 360-foot hill known officially as Devil's Mountain, and to Berliners as Mt. Junk or Rubble Hill. The hill offers ski jumps and toboggan courses for winter sports, has a military observation post on top, and even has a vineyard

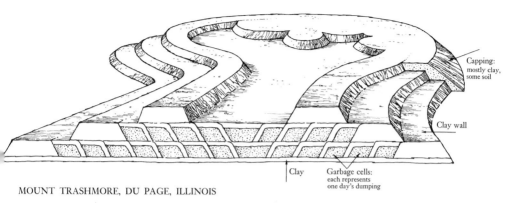

Capping:
mostly clay,
some soil

Clay wall

Clay   Garbage cells:
each represents
one day's dumping

MOUNT TRASHMORE, DU PAGE, ILLINOIS

Just west of Chicago, there is an old marshy clay pit which is
being excavated to form a lake, while the clay is used to build a
mountain of garbage, honeycombed into cells each four feet deep
—three feet of garbage, one foot of clay. Each layer is surrounded
by a clay wall. Clay is impermeable and thus prevents contamination of ground water. When the mountain is completed, it will be
125 feet high and will have six toboggan runs and six ski slopes.
The idea of Mount Trashmore was conceived by John Shaeffer,
now with the U.S. Corps of Engineers.

planted on its slopes. Both New York and Chicago are
considering building such refuse mountains to a height of
1,000 feet. Such mountains would cover five square miles
at the base and take 25 years to reach the proposed height.

Basically, however, there are three approaches to solid
waste which are ecologically sound. The first approach is to
cut down on the amount produced—less packaging; reuse
of containers such as bottles; use of cloth towels, napkins,
and handkerchiefs instead of paper; less production of
cheap, junky items which break down quickly and are cast
away. The second approach is to reverse the industrial
policies of planned obsolescence in durable goods—autos,
refrigerators, air conditioners, furniture, etc. There is no
objective reason why such items should not last a lifetime—
or several lifetimes. This would cut down production
enormously: instead of the 8 million new cars produced

each year, we might very well get along with 1 or 2 million. The third approach is to recycle used goods and refuse. The Salvage and Conversion System in Shawnee, Oklahoma, with the aid of Westinghouse Electric Corporation, is developing a system wherein a conveyor hopper takes all the collected refuse to a salvage center where marketable materials, cans, paper, plastics, glass, rubber, etc., are removed. The remaining matter is pulverized, composted with bacteria, and sold as soil conditioner.

Crucial to recycling is a good collection system of the solid waste, and New York City has experimented with some success on a Trash-Is-Cash program. The city has sites where cans, bottles, rags, and newspapers can be brought, and pays enough for the refuse to make it worthwhile. This requires a subsidy by the city—valuable junk, like copper scrap, is handled by junk dealers—but the cost is small compared to the benefits received.

An important development with regard to autos was made in late 1970. A monster machine was built, now operating in Detroit, which breaks up the bodies of cars (after engine and tires are removed) separates the various materials and shreds them or pounds them into the most suitable form for industrial reuse.

At the root of solid-waste pollution, as of all pollution, is consumerism—the drive for excessive possessions, excessive novelty, excessive variety, which has become a way of life in the United States. The fact is that the highest standard of living in the world, as hitherto defined, has resulted in the highest standard of garbage and pollution in the world. What we need is a new psychological approach: better, fewer things, but better.

● IO ●

FOR WHOM THE BELL TOLLS

Wherein we examine briefly noise pollution; the effects of noise. Noise as power. The tyranny of the decibels. The controversy over the SST *and the sonic boom. Some suggested remedies.*

> "No man is an Iland, intire of itselfe . . . any man's death diminishes me, because I am involved in Mankinde; and therefore never send to know for whom the bell tolls; it tolls for thee."

So wrote the poet John Donne, in the spelling of the sixteenth century, as he pondered on the mutual relationships of all men—a fine ecological viewpoint. The phrasing is peculiarly appropriate to noise pollution, for noise is always what the other fellow makes.

Sound pollution is on the increase; noise levels in cities have doubled in 16 years. Noise can be extremely harmful: in laboratory animals, noise exhausts adrenalin supply, shrinks lymph tissues, develops bleeding ulcers in stomach and intestines. When the body gets a danger signal from a loud noise, it automatically applies vasoconstriction—narrowing of the blood vessels. This hastens death, and at levels of 165 decibels, noise will kill animals such as rats and cats.*

In human beings, ear damage begins at 85 decibels. Loud rock-and-roll will reach 125 decibels and will cause deaf-

---

* A decibel is a unit devised to measure sound. The decibel scale is not arithmetical, but logarithmic, that is, 20 decibels are not twice as loud as 10, but 10 times louder. Normal human speech is about 50 decibels.

ness. The threshold is lower if the noise is constant. Some
other destructive levels are: truck traffic, 90 decibels; subways, 100; jackhammers, 110; jet planes at takeoff, 115.
Contrary to general opinion, people do not "get used to
noise." Noise is energy and doesn't get lost; it has impact,
real physical impact, not only on the ear drums, but on
other parts of the body; the skull, the torso and the groin.
Men wearing earmuffs while working around jet engines
develop diarrhea, nausea, and pneumothorax. Studies have
shown that workers in noisy surroundings are more argu-
mentative, show more signs of fatigue, and have more
neurotic complaints than other workers. There is evidence
that babies in the womb respond violently to noise during
the last three months of pregnancy.

In our steel and concrete cities machines have to be
powerful: jackhammers, bulldozers, air compressors, high-
impact rock drills, cement mixers, pile drivers, steam
shovels, compactors, riveting guns. Each seems to be noisier
than any of the others when it is working in your vicinity.
Add to these the recreational noisemakers: 450 million
radios sold in 20 years, 130 million TV sets, 83 million
phonographs, 33 million tape recorders. Then there are
trucks, cars, motorcycles, and the sirens of ambulances,
squad cars, and fire engines. The result is a steady, grinding,
never-ending background of noise that flares up suddenly
to ear-splitting levels.

One of the problems about noise noted by Robert Alex
Baron in his book, *The Tyranny of Noise* (St. Martin's
Press, 1970), is the fact that many people equate noise with
power. Not only is this the case with manufacturers of
construction equipment, but also with many ordinary individuals, from the motorcycle bug who feels the explo-

sions of his bike are mating calls (or overcompensation for sexual inadequacy) to the high-fi nut to whom sound is power. Salesmen play on this trait; Baron quotes a Fisher radio advertisement: "Power! Power! The power to unleash wattage and make an almighty noise is a favorite fantasy of the high-fi extremist . . . Take for example the devastatingly powerful Fisher system [which] can blast the roof off your house if you have an itchy volume-control finger."

The public has begun to be aware of the amount of harmful noise, and lawsuits are being instituted, particularly by people and communities adjacent to airports. In France, eleven suburban towns around the Orly airport combined to bring suit in July 1970 against Air France, Pan American, and Trans World Airlines. They are seeking payments for soundproofing 17 schools and 5 hospitals. Their suit charges that noise levels up to 114 decibels have been recorded in classrooms, and teachers claim that they lose an hour a day of teaching time waiting for the noise of takeoffs to subside so they can be heard.

In the United States the most dramatic struggle against sound pollution has centered around the sonic boom of the projected supersonic transport, the SST.

A sonic boom is the sharp, loud noise, very similar to an explosion, which is produced when a plane flies faster than sound, popularly called "breaking the sound barrier." This popular expression is misleading, because the plane doesn't break the barrier once, the noise is not created once, but is continuous throughout the flight. A sonic boom lays down a noise carpet of very high intensity that can shatter windows and crack plaster, walls, building foundations, tiles, and masonry. It can shatter fragile antiques and art objects in homes and museums. It can also trigger rock slides. In 1966 a boom from an Air Force plane caused 80,000 tons of rock to fall on ancient cliff dwellings in Arizona, causing irreparable damage. In 1968 a sonic boom loosened 66,000 tons of rock in Mesa Verde National Park in the Rocky Mountains.

Because of these experiences, ecologists have been up in arms at the idea of building a supersonic transport plane, the SST, designed to fly at 1,800 miles an hour. Its development is estimated to cost 2-to-3 billion dollars. Nearly all of it would come from federal funds—that is, taxpayers' money. Proponents of the SST argue that this money would be repaid when the plane is in production, but many observers are highly skeptical. The SST is an excellent example of private gain overcoming the public good, they say. Present jet planes fly at 600 miles per hour and thus take five or six hours for transatlantic flights. The SST would do the trip in two hours. But because of the normal delays in getting to and from airports, etc., it is estimated that the present 11 hours of door-to-door transatlantic flights would be reduced to eight hours, a gain of only 27 per cent to be measured against the cost, the discomfort, and the hazards of the SST.*

One of the major arguments used in favor of the SST is that both the French and the Russians have developed a supersonic transport plane designed for 1,400 miles an hour, and our national prestige is involved. But if those planes turn out to be uneconomical, as many observers believe, and are prohibited from flying over populated

---

* For a devastating analysis of the SST see *SST and Sonic Boom Handbook*, by William A. Shurcliff, Ballantine Books, New York, 1970. For the effects of sonic booms on sleep see the *Bulletin of Atomic Scientists*, May, 1970.

areas because of their sonic boom, then why copy them?

Soon after his election, President Nixon appointed a committee to review the problems of the SST, and most of the members recommended that the United States should stop developing the plane. However, on September 23, 1969, the President announced his intention to go ahead. Congressional opposition developed rapidly, fostered by citizens' conservation and ecological groups, and by the fall of 1970 it was clear that Mr. Nixon's plan was in trouble. Despite all pressures from the military and the airplane companies, on December 3, 1970, the Senate voted, 52–41, to reject the Administration's request for 290 million dollars for further development of the SST. Because the House had voted for the appropriation, a compromise was reached in the closing days of Congress to pass an interim appropriation of 210 million dollars until a vote could be taken by the Congress elected in November 1970.

In March 1971, the House and the Senate voted to discontinue federal funds for the SST—a remarkable tribute to the ecological consciousness which has developed in the United States.

As is the case in all pollution, sound pollution can be overcome. All machines can be made quieter, whether riveting guns, jackhammers, subways or airplanes. Submarines are potentially very noisy because of the heavy concentration of machinery—powerful diesel engines, compressors, pumps, generators, fans, etc.—but they must be quiet to avoid detection, and they are. Their machinery is designed to be quiet. The problem, as usual, is not technology but concern, money, codes, and enforcement.

The United States has lagged behind in the control of sound pollution. Europe has had antinoise codes since 1938, and these have extended into the Communist countries. In 1968, New York was the first American city to insert antinoise provisions in its building code. In 1969, Connecticut, which has recorded decibel levels higher than 94 on its turnpikes, began experimenting with a noise-indicator with a view to developing a noise control code.

As Robert Baron points out, the primary need at present is to get standards established, preferably nationally, and no sound-producing product would be allowed that exceeded those standards. In evaluating the noise emission of a given product, account would be taken of other noise emissions that would coexist with the product in question. It makes no sense to set a standard for a jackhammer, one for an aircompressor, and one for a truck or bulldozer, any one of which is low individually, but pile up to unacceptable levels when working together. Concurrently environmental standards could be set, guidelines for construction work, for traffic in cities, for airports, and so forth. We need to make quiet a public good, subject to public policy.

◗ II ◖

THE CITY AS AN ECOSYSTEM

Wherein we deal briefly with the urban environment. Overcrowding as a form of pollution. What a city should be. The size of cities and the difficulties of controlling them. Megalopolis and the urban sprawl. The example of Japan.

A city is as much of an ecosystem for man as a lake is for fish. It can be abused, polluted, and made unfit for man as surely as Lake Erie has been made unfit for fish. In the last seven chapters we have discussed major forms of pollution, each described separately for clarity, but all of them in fact inextricably intertwined. The totality of their impact is seen most clearly in urban living.

The degradation of cities is self-evident. They are slowly being paralyzed as autos choke the streets, requiring ever more space for parking and for transit; as the air becomes unbreatheable; as the water becomes undrinkable; as noise levels become unbearable; as the solid waste strains the resources of government. Add to this the overcrowding and bad housing in poor areas and the black ghettoes, the general ugliness, lack of parks and green areas in the cities, and it becomes clear that urban living has placed man under a constant physical and psychological stress. This is reflected in the statistics of crime, drug addiction, and mental illnesses.

Overcrowding is a form of pollution*—the jamming of

---

* So are poor housing, traffic jams, excessive and uncomfortable commuting, alienation from one's neighbors—any aspect of the environment which has an adverse impact on man. We don't go into them because of space and lack of facts on their consequences.

people in small apartments and dreary sprawls of tiny houses that make up the ugly outskirts of the city, really the bedrooms of the inner city. In animals, severe overcrowding has serious biological consequences, including madness, sterility, convulsions, and self-destruction. It is hardly likely that the human species is immune, and more research is needed to find out what does happen to human beings. Professor Paul Ehrlich of Stanford University has conducted experiments which show that crowding fosters aggressiveness in males.

As cities grow and become crowded, municipal services are adversely affected. Contrary to popular belief, per capita costs for services increase with greater concentrations of people. For example, telephone service costs more in a large city than in a small one because far more switching systems are required. When the number of people increases fourfold, the number of possible links between them multiplies 16 times. This means a growth of telephone exchanges, and since these exchanges are designed to handle only a fraction of the potential calls, a severe emergency, when everyone is trying to telephone, jams the exchanges and brings the system to a standstill. This happened in several cities during the great power blackout of 1967. Even without a severe emergency, the system can become irritatingly inadequate, as was the case with some New York exchanges throughout 1970. At peak hours in the morning, one had to wait as much as 10 minutes for a dial tone—and then pray the call would go through correctly.

For urban living to regain those normal amenities which give zest to life, not only must pollution in all forms be abated, but the city as a whole must be conceived of as an ecosystem. The city needs care, nurturing, a decent respect for the past, and foresight for the future. If beauty is part of man's needs, and few people deny that it is, then ugliness and uniformity are the enemies of the city. Narrow commercial interests make for uniformity, and they must be regulated. Zoning codes should be biased toward esthetics. Anyone who has seen central Paris, or Prague, or London knows that there can be an urban landscape as rewarding and satisfying as any countryside.

A city is a congeries of neighborhoods with differing developments and characteristics which should be preserved for variety and color. It needs unexpected turnings and corners, little parks and nooks as well as large parks and vistas to set off architectural gems; it needs fountains and sidewalk cafés. Walking in the city should be a pleasure, and walking is impossible when the automobile is king.

In the process of city planning, the automobile must be tamed. The automobile requires more parking space (136 square feet) than the average city dweller has for his bedroom. It is as contemptuous of people as a feudal lord, usurping grassy open commons for asphalted parking lots, contributing to noise and air pollution, causing sidewalks to be narrowed so roadways can be widened—all in all making the city dweller a second-class citizen compared to the auto.

What is the answer to the automobile? Essentially it is to make it difficult for the automobile to come into the inner city by building bypassing highways and by instituting tolls at entrance points, whether on bridges, tunnels, or highways. The money from the tolls should go into constant improvement of mass transit systems; the steeper the tolls, the more money for rapid transit systems, the more comfortable and attractive the system. Such a policy would discourage the use of the automobile, while encouraging the use of mass transportation.

A mass transit system is the spine of a city. Buses should be electric-powered, frequent, and cheap; subways should be clean, quiet, and the stations designed for charm and variety. It is a disgrace that New York should be so far behind Moscow in subway design, just as our railways are so far behind Japan's. The subways should tie into the big office, theater, and shopping complexes much as was done in New York with the Rockefeller Center. Both Chicago and San Francisco are planning their subways with this in mind.

The problems of cities are manifold; the solutions are controversial, and both are beyond the scope of this book. Many studies have been made, and there is a distinct swing in the United States toward planning and enhancing the inner city. Abatement of pollution of all kinds is the single most important contribution that can be made to improve the quality of urban life.

A more long-range goal should be to control the growth of cities and keep them down to a manageable size. There is no need or advantage in huge cities. One or two million people are more than enough to support all the amenities and variety of urban life. The larger the city the more difficult it is to administer it, and the 10-million cities such as New York, Tokyo, and London are virtually ungovernable—proliferating monsters that defy human ingenuity.

To control the size of cities is an extremely difficult task. In the Soviet Union, the government has made stringent rules, including a residence permit, to keep Moscow's population down, but the strict rules have been evaded and the city has steadily grown toward the seven-million mark. The size of cities cannot be controlled by decree. One way to cut down size is to develop more cities, and this, too, cannot be done by decree. It requires imagination and planning on a national scale, relocation of industries and services, decentralization of plants, skillful selection of locations for colleges, universities, research institutes, using of natural advantages, and so on. The United States is expected to grow by 100 million people in the next 30 years; imaginative planning now will provide model new cities and towns for these 100 million. In the process, we can remedy some of the worst of the present conditions, not only in the great commercial and industrial centers such as Chicago, but also in the purely industrial towns such as Detroit. Pollution control and dispersal of plants over a 20-year period can reshape urban life throughout the country.

Urbanization is here to stay, and the solution is not to return to a rural past which is gone forever. As in industry, the responsibility for the asphalt jungle lies in man's institutions and the priorities he sets himself. If the United States had decided in 1960 that a clean automobile or clean cities were more important than placing men on the moon, we would have a clean automobile today, while the moon would be explored by instrumented machines. A tenth of the ingenuity and money spent on manned spaceships would have given us not one, but several, experimental mass transit systems designed for speed, quiet, and comfort.

Urbanization is not only a matter of the great commercial and industrial centers. It is also a matter of the great urban sprawl, the running of communities one into the other so that there is one huge megalopolis from

WHO OWNS LAND

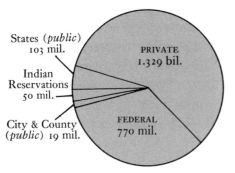

States (*public*) 103 mil.

Indian Reservations 50 mil.

City & County (*public*) 19 mil.

PRIVATE 1.329 bil.

FEDERAL 770 mil.

Total: 2.271 Billion Acres

HOW IS IT USED

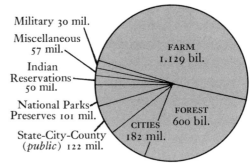

Military 30 mil.

Miscellaneous 57 mil.

Indian Reservations 50 mil.

National Parks Preserves 101 mil.

State-City-County (*public*) 122 mil.

FARM 1.129 bil.

CITIES 182 mil.

FOREST 600 bil.

Total: 2.271 Billion Acres

	1900	1940	1970
ACRES OF LAND PER PERSON	26.3	15.2	9.5
SHIFT OF POPULATION TO CITIES			
Total Population	76 mil.	132 mil.	204 mil. (*est.*)
Per Cent Urban Population	40%	59%	75%
Number of Cities Over 1 Million	4	10	25

Boston to Washington. Regional planning is essential to control such urbanization. A policy of drift will subvert beauty in city and country, as is happening to Japan, once renowned for the delicacy of its settings and the artistry of its people. A recent report by an editor of the *New York Times* is a dirge to a violated landscape:

> "[There] has been such extensive damage to the environment that life in the most heavily populated parts of these islands theatens to become unlivable . . . Of the 100 million Japanese . . . two-thirds live in cities, and of these the great majority are squeezed, at the highest population density in the world, into the narrow coastal plain that runs 400 miles south-westwards from Tokyo to Kobe and on to Hiroshima. These are the people who are existing in air so foul, on waters so polluted, in cities so choked and in 'slurbs' so ugly that by comparison the New York-Washington corridor looks almost like a planned development . . .

> "What seems especially tragic is that the Japanese have apparently learned nothing from the appalling experiences of Western Europe and the United States in sacrificing environment to the insatiable demands of economic expansion. And what doubles the tragedy is that this artistic and sensitive people could have been so indifferent to the need to preserve the very qualities of grace and beauty that have marked their unique civilization.

> "The ancient Imperial city of Kyoto, treasure house of Japanese art and architecture, is almost lost in a sea of smog. Along the fabled shores of the Inland Sea, petrochemical complexes spew forth long plumes of poisonous smoke across waters reputedly as polluted as Lake Erie. A famous shrine stares through its Shinto gate directly into a haunch of raw mountainside on the opposite shore, sliced away through the green hills as though by a giant butcher's knife.

> "If all this sounds like the United States, it is because it is like the United States—but the suddenness and intensity with which it has occurred make it infinitely worse.

> "If Japan is to be saved from an environmental Hiroshima, it is only the force of public opinion that will save it . . ."
> (John B. Oakes, *New York Times*, November 30, 1970.)

◐ 12 ◐

THE POPULATION EXPLOSION

Wherein we speak of population and the pollution and discover some interesting facts. The reality of the population explosion and its dangers. A key dilemma. Some suggested remedies.

That *Homo sapiens* is the great polluter has been amply shown in this book. So, one would think, the more people the more pollution, and this conclusion seems so obvious that, repeatedly, defenders of the *status quo* blame the increase in pollution on the increase in population. Some have even gone so far as to define people as pollutants. This is nonsense, for while all men are polluters, some are worse polluters than others, and an increase of population does not mean necessarily an increase in pollution. It depends on which population we are talking about.

Pollution is *not* the dumping of waste in the environment; it is the *excessive* dumping of degradable waste and the dumping of *nondegradable* waste so that the environment is overburdened and its recuperative powers overwhelmed. The waste of a Seneca Indian relieving himself in the Hudson River is not pollution; the domestic and industrial sewage of Albany, Troy, Newburgh, Poughkeepsie, and Yonkers *is*. The Seneca tribe could have quadrupled its population and not have affected the Hudson at all; doubling the population of today's river towns under present conditions would "kill" the river.

This principle is crucial to a study of world pollution.

One American pollutes the earth more than 100 Asians. The impact of industrialized agriculture is even worse: 3 million farmers in California produce, *through their farming methods*, a hundred times more pollution than all the teeming millions of farmers in India. The following table gives some pertinent figures:

	PERCENTAGE OF WORLD'S POPULATION	PERCENTAGE OF POLLUTION*
U.S.	6%	40%
Europe	13%	25%
U.S.S.R.	7%	15%
Japan	3%	5%
Rest of the world	71%	15%
	100%	100%

On a per capita basis, the United States pollutes the world three or four times more than other industrial areas and 33 times more than the underdeveloped areas.** It is abundantly clear that it is not the *size* of the population which is decisive in pollution, but the *kind* of pollution —the degree and rate of growth of its industrialization. A stationary population can easily contribute an ever-growing amount of pollution as is the case of Japan. In the decade 1958–68 Japan's population increased only 10 per cent; its per capita production (and pollution)

* It is difficult to be precise about pollution as a whole. I have constructed this table on the basis of per capita consumption of energy from all fuels as given in the *Statistical Abstract of the United States, 1970*, p. 821, with certain adjustments to compensate for density of population, advance of industry, etc. The 40 per cent I have used for the United States is a low estimate; most ecologists place it at 50 per cent and even 60 per cent.

** These are global figures and do not contradict the ratio of one American to 100 Asians previously given. The above table can be translated into showing how each 1 per cent of the world population living in each area contributes to world pollution. The figure for the U.S. is 6.5 per cent.

increased 400 per cent! Both rates are continuing, and unless drastic measures are taken, substantial areas of Japan will be literally unfit for man within a generation.

Industrialization is the primary cause of pollution; more precisely, industrialization as hitherto carried on throughout the world—in the U.S.A., the U.S.S.R., Japan, and Europe. What is urgently required, and what can be achieved, is industrialization with an ecological consciousness. Under present conditions, one half of the world population operating at U.S. levels would make the earth a wasteland; double the present population industrializing with an ecological orientation could live comfortably on earth with a standard of health, education, and culture in no way inferior to that of an American college professor circa 1970.

The argument I have presented so far is strenuously opposed by some population experts such as Dr. Paul Ehrlich, Professor of Biology at Stanford University and spiritual founder of the movement known as Zero Population Growth (zpg). At the annual meeting of the American Association for the Advancement of Science held in Chicago in December of 1970, Dr. Ehrlich and his supporters argued that population growth is responsible for pollution, for crime, and for the higher per capita municipal costs in California.

Opposing Dr. Ehrlich was Dr. Barry Commoner, Director of the Center for the Biology of Natural Systems at Washington University in St. Louis. Dr. Commoner and his supporters argued that with a properly designed technology, the United States could support a considerably larger population without damage to the environment. Dr. Commoner went so far as to say that the population problem was beyond the reach of scientists. Efforts should be concentrated on antipollution technology, and the population problem could take care of itself.

A few weeks later, Dr. Commoner's viewpoint received powerful backing from the chief official demographer of the United States, Conrad F. Taeuber, Assistant Director of the Census Bureau and a leading population expert. Speaking at Mt. Holyoke College, January 13, 1971, he underlined the difficulties of population forecasting. In the 1930's, expert opinion forecast a U.S. population of 153 million by 1980; that figure was reached in 1951, and by 1970 it was 206 million. "Time and again," he said, "American women—with the help of their male partners —have proven the forecasters in error." Women's attitude on number and spacing of children can change very rapidly, and the difference between having two or three children is enormous. With two children, the population is stationary, with three it grows very rapidly. He further agreed with the views on pollution expressed here:

> "Pollution, high crime rates, transportation problems and other social ills are not primarily a result of our population growth . . . Our population problem is one of tackling the agenda for improvement of our total environment. A lowered rate of population growth may facilitate the tackling of these tasks—but is is only one element in the programs which need to be developed."

As a matter of fact, in most of the United States population is *decreasing*. Rural America is becoming depopulated; villages in state after state are deserted. *Three quarters of all the counties* are "exporters" of people to the cities. In one-quarter, the emigration is barely made up by births so that the population is stationary; in one-half there was during the sixties a net drop of population despite births. The depopulation of the countryside is matched by the congestion in the cities. This process needs to be reversed through a national planning board for industry location. In this day and age, it makes no sense for each industry to decide unilaterally and without consultation where a new factory should rise.

On the issue of population and pollution, Dr. Ehrlich and the ZPG movement are, in the opinion of some people, exaggerating. Even if ZPG were adopted overnight and all women have henceforth only two children, the population of the United States would grow apace because of the very large number of young females who will reach maturity in the next 20 years. The population therefore could not become stationary until the year 2073 (Mr. Taeuber's estimate). The pollution crisis cannot go on that long; it must be defused within 10 or 20 years.

It should be emphasized that Dr. Ehrlich is *not* wrong about his concern with population growth, which is an extremely serious problem. Even if all pollution were eliminated from the earth, the population explosion would retain its full force as a threat to the human species, because in most of the world the food supply has not kept up with population growth. In the underdeveloped areas, malnutrition and starvation are rampant. The threat there comes *not* from pollution, but from famines which will trigger epidemics, plagues, revolutions, and wars of increasing intensity in the next 10 or 20 years.

Unwittingly, population experts who tie pollution to population instead of industrialization obscure the nature of the danger which the world faces ecologically and which we shall now examine in some detail.

What makes population growth so difficult to handle

is the speed of the increase, which is truly staggering and fully deserves the term "explosion." The rate of increase is measured by the number of years in which a population doubles—the so-called doubling time. A million or so years ago, according to educated guesses, there were 2.5 million humans in the world. By 6,000 B.C., the guess is 5 million. Doubling time, 1 million years. By 1650 world population was some 500 million people, giving a doubling time of 1,000 years. Next doubling time (1 billion in 1850) was 200 years, the next (2 billion in 1930) was 80 years, and the next doubling time 45 years (an estimated 4 billion in 1975). At the present time 1970, the population is 3.6 billion, and doubling time is 30 years, or 7 billion by the year 2000. Eighty per cent of this population will be in the underdeveloped areas—Asia, Africa, and Latin America.

Within recorded history, mankind has gone from a doubling time of 1,000 years to a doubling rate of 30 years. These figures are difficult to comprehend, but perhaps this comparison will have impact: from the time of recorded history, it took *8,000 years* to reach 1 billion people—it is now taking *eight years* to add a billion!

The present doubling time of 30 years is an average, ranging from well over 100 years for Austria to around 20 years for Venezuela. Imagine what it means to double the population in 20 years—to double the economy within one generation, double the food, the houses, the transportation, the fuel, the schools. It is a well-nigh impossible task without enormous outside help. Moreover these are current doubling times, but nearly half the population in underdeveloped countries is under 15 years of age, so that in the near future, *if nothing is done*, doubling time will go even lower.

These rates of increase must be cut sharply as quickly as possible, but this is easier said than done. The two basic determinants of population are the death rate and the birth rate—the number of deaths and the number of births per 1,000 people per year. Ignoring migration, the rate of increase or decrease is the difference between the two. If the death rate is 10 per 1,000 and the birth rate is 30, the rate of increase is 20 per 1,000 or 2 per cent a year. This gives a doubling rate of 35 years.*

Death rates throughout the world have fallen off sharply, as medical science has learned to control diseases and epidemics. But birth rates have not fallen in the same way throughout the world. In industrialized countries, they have fallen sharply; in underdeveloped countries they have remained high. The result is that the rate of increase in industrial countries is around 1 per cent; in underdeveloped countries it is around 3 per cent.

Why industrialized countries should have lower birth rates than underdeveloped countries is not fully understood, since the causes are many, subtle, and difficult to isolate. But some are obvious. In conditions of primitive agriculture, children cost little to raise, mean additional labor for the family and provide some security for old age. So they are welcome. In the United States children are expensive to raise, are a problem in urban housing, go

---

* A simple adding of 20 persons a year would take 50 years to add 1,000. But the 2 per cent applies each year to the previous addition (like compound interest in savings banks), so the time is shorter, as the following table shows:

ANNUAL PER CENT INCREASE	DOUBLING TIME
1	70
2	35
3	24
4	17

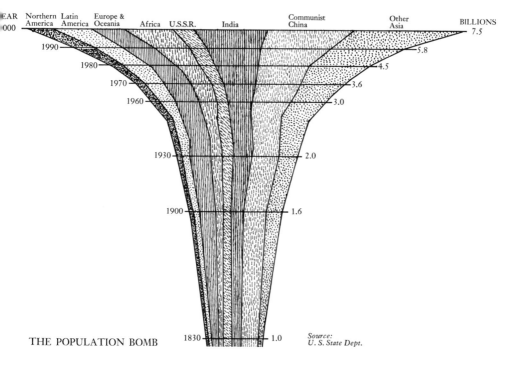

YEAR
2000

| | Northern America | Latin America | Europe & Oceania | Africa | U.S.S.R. | India | Communist China | Other Asia | BILLIONS |

THE POPULATION BOMB

1990 — 5.8
1980 — 4.5
1970 — 3.6
1960 — 3.0
1930 — 2.0
1900 — 1.6
1830 — 1.0

7.5

Source:
U. S. State Dept.

to work when grown up, contributing little to the family. Old age is partially insured through Social Security. Also, the American family, desiring birth control, has knowledge and means readily available. In India, ignorance is dominant. Customs play a part: in America we are not as set in doing things the old way as are the Hindus. Psychology plays a part: in many parts of the world having a lot of children is a mark of manhood. Where the woman has a degree of equality with man, she can have a say on how many children to bear; this is impossible where the woman is completely subservient. And so on.

Whatever the reasons, the fact is that industrialization leads to lower birth rates. Therefore, if we wish to

slow down population increases, industrialization seems the
most effective method. There is an important additional
reason for industrialization, and that is the desperate need
for food in underdeveloped countries. Industrialization
makes for more efficient farming and larger food supplies.
In the United States one farmer feeds himself and 30
other people; in Africa it takes eight peasants to feed
themselves and one additional person. This is a productivity ratio of 240 to one. Since some 70 per cent of the
world's population is in underdeveloped areas, it is there
that industrialization is urgently needed.

But because industrialization takes time and money—a
lot of money—many governments have tried to use massive birth control programs financed by the United Nations or developed countries. They found that widespread
and effective birth control is impossible without education
of the people, large numbers of doctors, nurses and administrators, good roads, good communications—in short,
all the qualities of a modern society—that is, an industrialized society. Japan, a modern society, instituted a
massive birth control program and within 10 years reached
one of the lowest rates of increase in the world, .8 per
cent. India, after great efforts at birth control, actually
went up in the rate of increase—from 2.3 per cent in
1958–63 to 2.5 per cent in 1963–68. While the birth rate
went down somewhat, the death rate went down even
faster.

Official birth control programs are useful and necessary,
but they are not a substitute for the intricate, many-sided
causes which industrialization sets in motion and result in
lower birth rates. But here we come to the fundamental
ecological dilemma of the world today: *if underdeveloped*

countries industrialize with current methods, such indus-
trialization will devastate the earth.

The dilemma seems inescapable: population increases must be slowed down or the world will explode in wars and revolutions. Industrialization will slow down the population increase and will ruin the earth ecologically. On one extreme, thermonuclear war; on the other, poisoned earth.

It is a grim alternative, and while somewhat exaggerated, not at all improbable. There is an answer—at least in theory: man must develop industrial techniques which are harmless to the environment. For the developed nations this is not an impossible task: both the technology and the financing are available. What is lacking is a popular will expressed in politics, and there is definitely a trend in that direction. For the underdeveloped countries, the problem is much more difficult, for not only are technology and finances at a low level, but time is crucial as the population keeps growing. The relation between costs and time is shown in a study by an Indian planner, Professor Mahalanobis, who considered the various alternatives for providing the 700,000 tons of grain required to feed the annual Indian increase of 5 million people. The grain could be supplied as follows:

a) Buy it from abroad. Cost $900 million; grain immediately available.

b) Buy enough fertilizer to increase crops. Cost $300 million; grain increase in one year.

c) Buy a fertilizer plant. Cost $200 million; grain available in five years.

d) Buy an engineering plant to build one fertilizer plant a year. Cost $28 million; time to take effect, 10 years.

Full industrialization would provide food at only 3 per cent of the cost of buying it from abroad. A proper world program would use all the alternatives but put the emphasis on the last one. The developed countries have to assume the major burdens of such a program and whether they will do so in time is perhaps the most crucial problem of world politics. Yet it can be done. The leading authority in the world on population and food was Lord Boyd-Orr who stated (*New York Times*, December 17, 1970):

> "If modern methods were applied in full to countries with food surpluses, the surpluses could be doubled. If applied in the food-deficit undeveloped countries the output could be doubled and redoubled . . . Food can be produced in abundance and the explosion of population halted if all governments were effective and willing to apply the necessary measures.
>
> If a permanent World Food Council, including representatives from food-deficit countries, could be set up to deal with the world food problem, agricultural and engineering technology could eliminate hunger from the world within a decade and our civilization not be threatened by a world food shortage."

Part III

MAN THE CREATOR

The preceding chapters, detailed though they are, give only the bare bones of the problem of pollution. It can only be solved by conscious social measures which are bound to make many people uncomfortable and to conflict with many institutions and interests.

An easy way out of the conflict is to say that technology can solve the problem of pollution. This is a half-truth. Technology can solve the problems of pollution *provided that it is used as a tool in the hands of socially minded and socially conscious citizens.*

The most promising technological advances are in the field of recycling. But that is only part of the measures needed. Some cuts in consumption (as in packaging, automobiles, energy) are also necessary, and to carry those out, as well as to foster technological measures, the society must take the proper economic and sociological measures. The motor is the pressure of citizens—politics. In the most fundamental sense, ecology is politics.

● 13 ●

TECHNOLOGY AS SERVANT OF MAN

Wherein we talk of men and machines. Then of food and energy. If not DDT, *what? If not automobiles, what? If not oil wells, what? Some suggestions, and a brief flight of fancy to the year 2001.*

From the beginning of the Industrial Revolution in the eighteenth century, men were suspicious of machines. The first and most obvious fear was that of being displaced, and in 1811 a group of English workers, known to history as Luddites,* systematically wrecked new textile machinery which they felt was responsible for their unemployment. As the centuries went by and machines developed, becoming ever more sophisticated and complicated, some people began to fear that technology was dominating man, shaping his thought and mechanizing his emotions. The development of computers, so fast and free of error, has given rise to the idea that they will supplant man, a theme dramatized in the film *2001: A Space Odyssey.*

The fact is that computers are programmed by men and operate on data given by men. Experts have a saying, "garbage in, garbage out" (GIGO for short), to underline this truism. Man uses tools; tools do not use men. In regard to man's goals and desires, tools are neutral, and there is no such thing as an evil instrument or a saintly instrument. The sharp knife in a psychopath's hand is

---

* The word is derived from a certain Ned Lud who, around 1779, smashed two machines of his employer. It is not irrelevant to note that Lud was feebleminded.

a murderous instrument; in the hands of a surgeon it is a healing instrument. But the murder and the healing are in the mind and heart of the man. Technology has never been, and technology can never be, the master of man.

In our generation, science and technology have expanded so enormously that men, feeling helpless and in despair, turn to a kind of intellectual Luddism. They wish, for example, that nuclear technology did not exist. But it is not the machines that are responsible for the alienation of modern man; it is the social institutions. What mankind needs is not less nuclear knowledge and technology, but more. We need to achieve controlled thermonuclear fusion which would be the answer to almost all of our pollution problems. We also need to control the men and institutions which stockpile nuclear weapons. The two are not incompatible.

The proper use of technology depends on knowledge. One of the great developments of the 1960's was the awareness of the academic world of ecological problems, the awareness that ecology is *par excellence* an interdisciplinary science and that special arrangements must be made within the universities so that it can be taught without interdepartmental frictions. Many universities now have special departments for ecology and environmental studies.

One of the more exciting academic developments was reported by John Fischer in *Harper's* for February, 1971. This is the establishment of what Fischer calls a "survival university," where all work will be focused on a single unifying idea, the study of human ecology and the building of an environment where our species might survive. Such a university was opened in 1969 in Wisconsin where a cluster of four campuses located in and around Green Bay have been organized as a special kind of ecological university. The president of Green Bay is a political scientist, Edward W. Weidner, who is breaking ground by discarding the usual departmental setup in teaching. The university is organized around four "theme colleges" and one school of professional studies, each granting its own kind of degree. The four colleges are: College of Environmental Sciences, College of Community Sciences, College of Creative Communication, and College of Human Biology. A student chooses a theme (say regional planning for the Susquehanna Valley) and gets his degree in that, choosing whatever courses he needs to achieve his aim. At Green Bay, it is basic that technology is the servant of man.

Once this principle is grasped, we can focus on technology's potential. We have seen, for instance, in Chapter 4, how the tertiary plant in Lake Tahoe solved its pollution problem and how the Muskegon County project may provide a solution for cities. There is another example of a different technical approach which has worked in Santi, California. There the effluent was piped to a holding pond on a gravelly hillside whence it filtered down to the valley to form eight irrigation ponds.

It may be stated flatly that the technology is available to eliminate practically all water pollution in the United States. What is needed is money: 10 billion dollars a year for 10 years to build storm sewers, the expansion of techniques to bring effluent to the land where possible, and the addition of tertiary stages to plants in places where soil conditions preclude the Muskegon plan.

As we saw in Chapter 7, the same flat statement can

be made about air pollution except for the case of the automobile. The Seversky Hydro Precipitol, the Ironton preheater, and many other devices can be installed. The cost of cleaning up air pollution is much lower than that for water pollution, perhaps a tenth or even less. The automobile is a more recalcitrant problem, and new types of engines must be developed and put into production. There is no doubt that a practical electric car will be developed within a decade by both Japan and Germany. The largest electric power company in West Germany announced on April 6, 1971 that it planned to have a battery driven car capable of fifty miles an hour in service by 1980 as well as an electric bus, together with a network of service stations. Two prototypes, a car and a bus, are currently being tested. The United States can only lag behind at the risk of losing part of its market, as it did in regard to the small car during the 1960's.

A gas turbine engine, remarkably free of pollution, is already in existence, and New York City has hired one of the foremost auto engineers in Detroit to build an experimental car with such an engine by the end of 1971. The engineer, Sam Williams, is also working on a gas turbine engine for General Motors. Mr. Williams told the *New York Times* (January 9, 1971) that in his opinion the auto industry would not be able to meet the standards set by the Muskie Bill. Around 1975, he thinks, the auto industry will have to decide to switch to the gas turbine engine. His judgment is indirectly confirmed by a speech of Henry Ford II in Detroit, May 27, 1971, asserting that it would be "an impossible task" to meet the standards in the Muskie Bill and questioning the validity of the air quality standards.

Another type of engine, the so-called Sterling engine, which has hitherto been used in refrigerators, is already being used in experimental buses in Eindhoven, Holland, built by the Phillips Lamp Company, one of the largest enterprises in Holland. The company announced (*N.Y. Times*, December 21, 1970) that it had perfected a smaller engine for cars and that it had licensed General Motors to use and further develop the engine.

It is highly probable that by 1980 there will be several types of cars in production, geared to different uses, all with clean engines. The auto companies have resisted the change-over because of the cost—an estimated 8 billion dollars. With government subsidies, and 10 years to make the shift, the economic cost should not be prohibitive.

An important recent development in antipollution technology is the awareness that it is often easier and cheaper to prevent pollution than to treat its manifestations. This is particularly true in industry. For example, aluminum is made by electrolyzing bauxite in molten fluoride. Fluoride is extremely toxic and quite a bit is discharged in the process, which is eliminated by powerful ventilation. The workers are protected, but the area around the plant becomes a wasteland. Although fluoride is valuable, it doesn't pay to extract it from the gas discharges.

But in the most recent techniques the electrolyzing baths are hermetically sealed and the fluoride gas is taken directly from the baths with little extra expense. Result: no pollution, no expensive ventilation, and no damage to the environment.

The value of waste products often pays for the whole of the antipollution efforts. In Russia, the Gorlovka nitrogen fertilizer plant in the Donbas region used to discharge annually 2,000 tons of sulphuric acid, 900 tons or nitrogen, 700 tons of ammonium nitrate, and 1,000 tons of ammonium hydrate—the output of a chemical plant. This waste was discharged into a gully and thence laid waste the surrounding vegetation.

Instead of designing expensive purification machinery, the chief chemist at the plant decided that the pollution should be stopped at the source in the production process, either by stopping it completely by new techniques (as was done for fluoride in the bauxite plant) or by recycling waste liquids as they appeared in the production process. This meant a much more complex manufacturing plant, but experience showed that one half of the chemicals recovered were so profitable that they subsidized the extraction of less profitable chemicals. On balance, the plant was made pollution-free at no extra cost to the enterprise as a whole.

The late 1960's saw a great development of applied technology for antipollution control, particularly in the area of detection instruments. For example, one reason why mercury pollution was not discovered earlier, before sickness gave the alert, was that there were no instruments sensitive enough to detect the small quantities of mercury in the water. Within a year after mercury contamination became an issue, several companies developed sensors that could do the detecting job. Sometimes it is not a question of inventing new instruments, but of applying existing instruments to antipollution work. A striking example is the use of infrared sensors for spotting water pollution. In December, 1970, New York City tried a successful experiment in which an infrared sensor was mounted on a helicopter which then crisscrossed over the waters in the metropolitan area. The sensor used could determine the difference in water temperatures as small as one tenth of a degree.

Aside from the spotting of obvious thermal pollution, such slight differences in temperature can point to polluting fluids going into a river, because such fluids, going through different areas such as a factory or a sewer, would have differing temperatures.

Perhaps the most exciting application of infrared photography and other sensing systems is through the instrumentation of unmanned earth satellites envisaged in the current Earth Resources Observation System (EROS). The first EROS satellite is scheduled to be launched in 1972 to reveal "ground truth," a phrase invented by photointerpreters to point up the difference between what infrared photography sees and what the naked eye sees. For example, a dying elm tree and a healthy elm tree look the same to the tree surgeon. But in infrared photography the dying tree shows a different color than the healthy tree. Since beetles lay their eggs only in dead trees, the timely removal of a dying elm can slow down or halt the spread of beetles.

The possibilities of EROS stagger the imagination. In an article in the *Saturday Review*, April 3, 1971, science editor John Lear gives a glimpse of what "ground truth" entails based "on the assumption that every kind of rock, every variant of soil, every family of tree and vegetable and moss and shrub, has an individual signature exclusively its own. Wheat signs differently from barley, barley differently from corn, corn differently from rice. . . . Once the identity is determined, the signature can be picked surely out of a multitude of other signatures, recorded, stored and run through computers—all automatically. A bank for preservation of such automatically prepared documents is to be constructed as part of the Earth Resources Observation System."

The Department of the Interior will spend 4.8 million dollars for the building and the supporting facilities to process 50,000 images a year. Such a bank can be developed in other countries; some three dozen countries have already expressed interest in the project which is to be dicussed at a U.N. conference in Stockholm in 1972.

Another use of infrared sensors is in determining the source of oil pollution in the ocean. Since oceanic oil pollution is potentially so cataclysmic, and since the national and geographical sources are so varied and dispersed, international controls are essential, but they can only be effective if the source is pinpointed. All over the world scientists are working on this problem and, in April 1971, the national meeting of the American Chemical Society devoted a substantial part of its proceedings to methods of detection. Four major approaches have proved fruitful: infrared spectroscopy, neutron activation analysis, mass spectrometry and gas chromatography. These are highly technical methods too complicated to explain here but they are all designed to "fingerprint" an oil slick and determine its origin.

The potential applications of technology to antipollution are so varied and numerous that it would take several volumes to discuss them. We will deal here only with two areas, agriculture and energy production, because they are of key importance.

The major problems of pollution in agriculture stem from the use of artificial fertilizers and of pesticides. The best solution to the fertilizer problem is the return of sewage effluent to the soil, as in the Muskegon County project described in Chapter 4. In this connection, one of the great sources of natural fertilizers lies in cattle manure which now accumulates in feed lots near big cities and is dumped into their sewage systems. A decentralization of packing plants would scatter those feed lots far and wide among farms which could then use the manure as part of an irrigating effluent.

The other problem in agriculture, the use of pesticides and herbicides, has been solved theoretically and is in the process of being solved practically. The solution is twofold: find a chemical that attacks only a given plant or insect and is harmless to others, and/or shift from chemical to biological controls which are also geared to a single species. Instead of spraying Japanese beetles, for example, you introduce a wasp that feeds on their young. Or you introduce a disease that attacks the beetle and nothing else.

The division between chemical and biological is not always clear, for often a biological method becomes chemical in application. For example, it has been discovered that mosquitoes, when too crowded together secrete a substance that kills their young. Scientists are working on a way to isolate that substance, reproduce it synthetically —that is, chemically—and use the chemical to spray the larvae of the mosquitoes. Another example is the use of sexual attractants. Some insects, responding to an odor emanating from the opposite sex, will come out of hibernation, or will fly to the source. By isolating and duplicating the chemical that gives forth the odor, insects can be lured to their destruction. How such an attractant works is illustrated in the case of the gypsy moth, which

attacks trees and foliage during its caterpillar stage and
has been extremely destructive in New England.

The female gypsy moth matures in July-August and
sends forth a strong odor which is picked up by the male
moth who follows the odor trail to his mate. After mating,
the female lays about 400 eggs, which hatch the following
spring.

In the early 1960's, the sex attractant was obtained from
the abdomens of 80,000 moths and used in traps to lure
the male, but the experiment was not very successful because the attractant was not very powerful. In 1970, the
Department of Agriculture succeeded in synthesizing the
attractant and the synthetic chemical turned out to be
30 times more powerful than the natural extract, both in
range and lasting qualities. The days of the gypsy moth
are numbered. Meanwhile work is progressing on a sex
attractant for that formidable pest, the cockroach.

The shift to biological controls is shown by the shift
in research of the United States Department of Agriculture. The entomological (insect studying) division of its
research center devotes 51 per cent of its budget to biological control methods compared to 16 per cent for
conventional insecticides (compared to 65 per cent 15
years age). Its director, Dr. Edward F. Knipling, is
credited with originating one of the most brilliant ideas
in biological controls—the eradication of an insect species
by introducing sterile insects (not able to reproduce) into
any given population of pests. The way this works gives
a fascinating glimpse into the problems of insect control
and extermination.

For our study we will take the boll weevil, which
destroys cotton. It is so destructive that 35 per cent of

all insecticides (mostly DDT) used in the United States are
directed to this single pest. Although it does not reproduce
as fast as the housefly, it does well enough. Its generation
span is 21 days, and a pair may produce up to two dozen
offspring. This means that a boll-weevil community which
has been 98 per cent destroyed will be back to its original
size within a year. What insecticides have done is to hold
the boll weevil in check by repeated sprayings several times
a season, year after year. Now let us see how sterilization
works. This example is taken from an article by Hal Hig-
don in the *New York Times Magazine*, January 11, 1970.

Assume that you have a thousand acres of cotton with
200 boll weevils per acre. This gives a population of
200,000 insects, which we want to exterminate completely
by current DDT techniques. It has been found that it takes
seven applications of DDT to reduce the population by
98 per cent, leaving 4,000 survivors. Since each pair will
have, on the average, 20 offspring, the next generation
(21 days later) will show a total population of 40,000
insects, a tenfold increase. Repeat this process and you
get the following table:

BEFORE DDT	AFTER DDT	21 DAYS LATER
200,000	4,000	40,000
40,000	800	8,000
8,000	160	1,600
1,600	32	320
320	6	60
60	0	

Complete extermination is accomplished, in theory, with
42 sprayings of DDT over a period of 105 days. The cost
would be $42,000, and every living thing in the fields

would be dead—a solution prohibitive both in dollars and in ecological consequences.

Now let us see what happens if we introduce 100 sterile insects for every normal one in the population, reducing the chances of reproduction by 100-to-1. Using the same example as above, it is clear that to introduce 20 million sterile boll weevils (100 times 200,000) would be to destroy the entire crop. Therefore, the first step is to spray seven times with DDT and reduce the population to 4,000. Then you release 400,000 sterile boll weevils and accept the damage to the crop. The surviving 4,000 will only reproduce 40 normal boll weevils—one for every hundred matings. You then release another 400,000 sterile insects, and here is what happens:

NORMAL WEEVILS	STERILE WEEVILS	CHANCES OF REPRODUCTION	SURVIVORS
4,000	400,000	100-to-1	40
400	400,000	1,000-to-1	0

Cost: $11,000 and only seven applications of DDT.

But even this picture can be improved if a sex attractant can be developed to lure weevils onto a sticky substance to be killed. Dr. Knipling thinks one might get an 80 per cent kill. Suppose, then, that at the end of the autumn you attack the 200,000 population with DDT, leaving 4,000 to hibernate. In the spring you could greet them with the attractant and get rid of 80 per cent leaving only 800 weevils to be overwhelmed by a single release of 400,000 sterile insects, thus cutting down on cost and the destruction of crops.

There are still technical problems before a concerted

attack on the boll weevil can be mounted, but the techniques of sterilization have been tested successfully against the screwworm fly in the island of Curaçao in the Caribbean, in Florida, and in Texas. The screwworm fly lays its eggs in open cuts of animals (cuts made from barbed wire, thorns, etc.). The eggs hatch into tiny maggots which feed on the flesh, causing infection and death. Chiefly the fly affects cattle, sheep, goats, and wildlife, but there have been instances of its laying eggs in the nostrils of sleeping humans with the result of the larvae reaching the sinus, infecting it, and even causing death.

In Florida, the screwworm fly has been successfully contained after an 18-month program using three billion sterile flies. At the present time, the United States government maintains a 300-mile buffer zone with Mexico into which 175 million sterile flies are released each week. With the cooperation of Mexico, a program has been begun to move farther and farther south until the screwworm fly is bottlenecked at the Panama Canal. It may take several decades, but the screwworm fly will eventually become extinct.

At present, sterilization is done by x-ray or other radioactive substances, such as cobalt-60. It is an expensive and time-consuming process. The search is on for a chemical capable of sterilizing insects which would speed up the process. Meanwhile all kinds of techniques are being researched. Hormone overdoses can create malformation and abnormal development resulting in death. By the use of artificial light or a sex attractant, insects are lured out of hibernation to freeze while it is still cold, or to starve before the crops are out. There is a suspicion that some insects are attracted to each other and to certain crops by infrared radiation. If this proves true they may be trapped by duplicating the radiation that attracts them.

The search for parasites and predators goes on. Recently, research scientists located a flea beetle in South America which feeds on the alligator weeds that choke inland waterways. The flea beetle attacks only such weeds and dies after its food supply is exhausted—a perfect control device. In Antibes, France, scientists are using ladybugs to control plant lice. One particular strain of ladybug is so voracious that one bug eats 100 plant lice a day. Recently, 600 of those ladybugs were flown from France to Mauritania where an oasis of some 6,000 date palms were being attacked by cochineal insects. The 600 ladybugs wiped out the insects. In effect, each ladybug—with an assist from its offspring—saved 10 trees.

Another line of attack against pests is through disease, because it seems that diseases that attack insects do not attack man. Scientists have already developed a virus that attacks gypsy moths, so that between this virus and the sex-attractant traps previously described, the gypsy moth should be under control in the near future. However, much more research has to be done on the use of diseases and their possible consequences on other insects and animals.

Perhaps a hundred insect pests cause 90 per cent of all damage. Twenty of these pests account for most of the damage. Dr. Knipling believes these 20 pests can be systematically eliminated at the rate of one a year. A final point to be noted is that he is operating on a research budget of 18 million dollars, a very small sum in the government finances—about the cost of five hours of war in Vietnam. Doubling the research funds would undoubtedly speed up the process of finding alternatives to pesticides and herbicides.

Equal only to the production of food is the need for the production of energy, whether as heat or as electric power. To control pollution due to energy production requires a massive research program to develop new sources of energy, such as harnessing solar radiation, the power of the winds, the ebb and flow of tides, and the difference in heat levels between the warm surface and the cold depths of the seas. In every one of these areas, pilot plants and experiments have successfully produced electric power, but the cost has not been competitive with oil, coal, and nuclear fuels. However, like the electric car, one reason they are not competitive is because research money has been stinted. Moreover, many of the proposed new sources of energy tend to be local: where tides are available, where winds are reliable, etc. Fossil-fueled and nuclear plants can be placed anywhere, will perform equally everywhere, and there is a bias in favor of that which can be widely standardized with a high expectation of regular performance. However, as the energy crunch tightens, local energy sources become more important.

The most promising source is the hot steam and water trapped under great pressure inside the crust of the earth several thousand feet below the surface. This so-called "geothermal" energy may be of great significance in certain areas such as western Mexico and the United States. The Mexican government has set up a Geothermal Energy Commission which has already drilled 17 wells to tap the steam underground. Most of the wells have been capped, but some are blowing steam for equipment tests while a major power plant is being constructed. The plant is scheduled to go into operation in 1972. (*New York Times,* August 22, 1970.)

Both Mexico and the United States already have experimental plants functioning, while Italy and Japan are also pressing research in this field. One of the problems within the United States is the fact that most of the geothermal steam lies beneath federally owned land in the western states—some 1,350,000 acres of land—and there is no precedent for leasing steam. The *Wall Street Journal,* (December 10, 1970), reporting at length on this new energy source, expects legislation to be passed in 1971 spelling out leasing regulations.

Geothermal energy is restricted to suitable areas and cannot provide too big a share of the world's energy requirements—at best, perhaps some 5 per cent. But it could be of strategic importance in a specific place, as for example, in Los Angeles, where conditions make it imperative that the air be completely pollution free.

Another source of pollution-free energy is natural gas, which burns almost completely, and hence has little polluting residue. Vast gas fields exist in Holland, Siberia, the Sahara, the American Southwest, and in Alaska. More will be found as exploration increases. The Soviet Union has built some 4,000 miles of large-diameter pipelines (40 inches and up), and is supplying Austria, Germany, and Italy. Present agreements call for the annual supply of 100 billion cubic feet to Western Europe by the late 1970's. The Soviet Union is also negotiating with Japan to build a 1,000-mile pipeline from Siberia to the Pacific Coast. In the United States, the network of pipelines is constantly increasing and runs into tens of thousands of miles. The north slope of Alaska will supply a large amount of gas, and gas pipelines do not have the ecological danger of spillage and heat the proposed oil pipeline would have.

Natural gas can make a substantial contribution to the world's energy requirements, and it merits government subsidies for a more rapid expansion of pipelines and exploration for new fields. Moreover, recent technical developments allow it to be liquefied under pressure and thus to be shipped in tankers. International trade in liquefied natural gas (LNG) is only six years old and is dependent on huge tankers that carry a billion cubic feet at a time. There are only 11 such tankers in existence now; there will be at least 25 more by 1975, and perhaps as many as 70 by 1980. These tankers do not involve the risks of spillage that oil tankers do, because if there is a collision the pressure is removed, the liquid becomes gas, and the gas is harmlessly dispersed in air and water.

LNG cannot contribute significantly to the world energy requirements as far as volume is concerned. A fleet of 100 tankers could only transport in one trip the equivalent of 1 per cent of all the fuel used annually by trucks and autos in the United States. But as a supplement to other sources of energy in coastal areas, it can be of great benefit. It can be used by industry and utilities to meet sudden, unexpected demands; it can be sent to isolated spots; it is an ideal emergency fuel for areas struck by catastrophe: hurricanes, tidal waves, earthquakes.

The ultimate solution to energy-production pollution lies in the development of a controlled thermonuclear reaction—controlled fusion. The raw material of nuclear fusion is sea water—unlimited (for this purpose), free, available to all regions of the earth, and nearly all nations. It is pollution free and should solve all energy problems.

MAGNETIC "BOTTLE"
EXTREMELY SIMPLIFIED

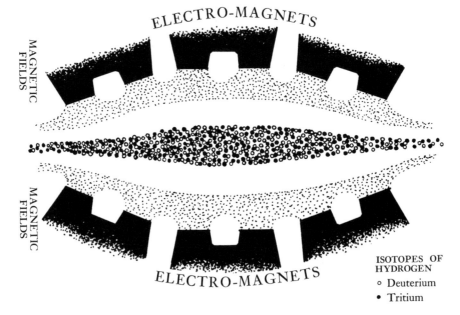

ELECTRO-MAGNETS

MAGNETIC FIELDS

MAGNETIC FIELDS

ELECTRO-MAGNETS

ISOTOPES OF
HYDROGEN
o Deuterium
• Tritium

Thus far nuclear fusion has been achieved only in an un-controlled state—the thermonuclear bomb.

The problem in the fusion reaction is the enormous heat and pressure required to fuse deuterium and tritium (two isotopes of hydrogen). The heat required is comparable to the heat of the sun's interior, and in the hydrogen bomb it is achieved by placing a *fission* bomb inside the fusion bomb. As the fission bomb explodes it generates the heat and pressure which produce fusion.

No known material could hold such heat: anything and everything would instantaneously vaporize. But scientists are resourceful and they developed a "magnetic bottle." The plasma (the gas to be fused) is held in place by power-ful magnetic fields so that it never touches the walls of the container and is then heated by electric impulses to enor-mous temperatures. All kinds of tricky problems developed in research, problems so complex and fundamental that countries were forced to abandon their secrecy in research and leave all the scientists of the world free to work together and discuss the problems with each other. English and American scientists have made large contributions to the problem of fusion, and in 1969 Soviet scientists achieved what was considered a major breakthrough on some fundamental problems.

Meanwhile other scientists sought other solutions, of which the most promising seems to be the use of lasers instead of the "magnetic bottle." By 1970 this approach seemed to be ahead.

Lasers are directed beams of light of a single wavelength which can be built up to tremendous energies and generate tremendous heat. The new technique is shown in the diagram on page 186. It consists of creating an empty cavity by a whirlpool of lithium into which are dropped frozen pellets of deuterium and tritium. A laser beam hits the pellets and vaporizes them, heating them to the temperature required for fusion. The pellets do not touch the walls of the container at all. The heat is absorbed by the molten lithium which carries it off to a heat exchanger where it is funneled out to make electricity. The scheme is theoretical, and more powerful lasers are needed before experiments can be made to see if the fusion can be controlled. But this example gives an idea of how some seemingly impossible problems can be solved.

A controlled fusion process may be achieved in the laboratory within the next 20 years in the judgment of a committee of distinguished scientists. The committee was set up by the American National Academy of Science to assess the kind and timing of major technological develop-

Harnessing the Power of the Hydrogen Bomb

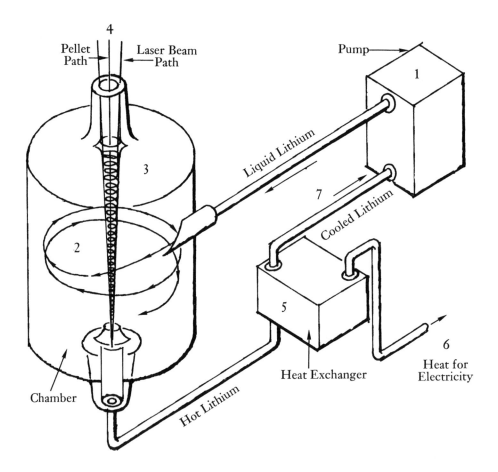

4
Pellet Path — Laser Beam ←Path
Pump
1
Liquid Lithium
3
7
Cooled Lithium
2
5
Hot Lithium
Heat Exchanger
6
Heat for Electricity
Chamber

(1) Liquid lithium is pumped into a spinning chamber, (2), where it whirls at high speed creating a cavity, (3). Frozen hydrogen pellets (deuterium and tritium) are dropped into the cavity while a strong laser beam vaporizes them, (4), heating the gas to fusion temperatures. The liquid lithium absorbs the heat given off in fusion and carries it off to a heat exchanger, (5), where the heat is sent off to create electric energy, (6), while the cooled lithium is recirculated, (7).

ments in the coming decades (*Scientific American*, February, 1970). From laboratory to industrial use is not an easy step. It may prove enormously costly in capital equipment, but since the raw materials are free it doesn't seem likely that the costs will be beyond the capacity of the advanced industrial nations—even if it should involve a consortium of them all, transcending ideology and economic systems.

Controlled fusion would be the supreme triumph of science and technology at the service of man. It may well achieve the ultimate dream of prophets, the race of man united in peace under a federalized world government. It could even happen by the year 2001 under the twin drives of a stick and a carrot—fear of a thermonuclear war and the alluring benefits of nuclear fusion. Some carrot, some stick!

Yet better to dream than to despair, and as George Bernard Shaw once said to a young man: "Be careful what you wish for: you may get it."

• 14 •

ECOLOGY AS POLITICS

Wherein we talk some politics and economics. The use of customs, the use of laws, the use of taxes. A choice of social priorities; what should be done first. What people can do, individually and in groups.

The thrust of ecology today is not only to stop harming the environment, but also to repair the damage already done. Ecologists are social doctors, so to speak, and just as a medical doctor cannot force his patients to get well, ecologists cannot force society to accept their cures. Hence, in the first instance, the problem is one of education—educating the ordinary citizens, the public officials, the industrial managers and policy makers. Through education, habits and customs can be changed. Not so very long ago thieves had their hands cut off; in England, one hundred-odd crimes carried the death penalty. Through education, society learned that state violence is no solution to crime.

A goodly amount of pollution can be eliminated by education. People can learn to save water and save electricity; they can learn to walk a little more and use the automobile a little less; they can learn to do with older clothes, older furniture, older stoves and refrigerators. In fact, a vogue could be developed that it is chic and sophisticated to buy fewer and more lasting goods.

But there is a limit to what the individual can do. He cannot initiate research for an electric car or test food for purity or the river for pollution. What he can do, however, is to join other like-minded people to see that his local, state, or federal government does the things they agree are necessary—through laws, regulations, and the power of taxation. Laws may be as simple as a municipal ordinance prohibiting spitting in subways, or as complex as a federal law prohibiting the use of pesticides.

In a democracy, legislation is the arena where conflicting interests clash and get adjudicated. In ecological problems there are usually conflicts not only among private interests but also between such interests and the public good. This holds for national questions such as the trans-Alaskan oil line, as well as local questions such as the pollution by a chemical plant in a small city like Stamford, Connecticut. In Alaska it is not only the faraway oil companies who are reaching for gain, but the community as a whole—the land developers, the construction companies, the stores, the service industries, the workers and their families. In Stamford, it is not only the management which resists the expense of abatement equipment, but the 200 workers of the factory who fear for their jobs and shower city inspectors with bricks and bottles. It is easy to blame companies or workers for their attitudes, but when one's profits or livelihood are at stake, civic responsibility fades away.

Legislators, national and local, are supposed to defend the public good, and will do so if citizens are aroused. A prime example is the legal control of auto pollution which was initiated by California in 1961. It culminated in the federal Muskie Bill of 1970 which requires that automobiles shall achieve a 90 per cent reduction of emissions by 1976. The auto companies bitterly fought the bill, will undoubtedly contest its provisions in the courts, and will probably get extensions in 1975, but the pressure is on, and already one hears of increased experiments with new types of engines discussed in the previous chapter.

The Muskie Bill also requires that newly constructed sources of pollution, such as power plants, paper mills, steel mills, cement mills and the like shall use the latest anti-pollution technology. Also passed by Congress in 1970 was a law tightening standards for pollution by oil and sewage at sea, by discharges from mines, and for thermal pollution from atomic power plants.

Federal legislation both follows and stimulates state and local legislation. In 1970, for example, the state of Illinois passed a pollution control act, which was described as the most comprehensive in the nation at a meeting of the 50 state attorneys general held in St. Charles, Illinois, June 29, 1970. The act abolishes special exemptions granted to polluters by localities, it forces a polluting corporation to post a performance bond to insure compliance, it empowers the Environmental Protection Agency to seal up offending equipment, it empowers the attorney general to institute environmental suits, and it eliminates legal challenges to his powers to prosecute violators. Illinois has the most effective legislation of the 17 states which enacted important environmental measures in 1970.

Municipalities are stirring. Both Seattle, Washington, and San Diego, California, cleaned up their bays in 1970 by a combination of expenditures and preventive laws. New York City tightened up its air-pollution code. Baltimore has passed a stringent air-pollution code, reputedly the best in the nation; San Francisco has passed regulations on pollution in its bay which the city claims are the most advanced in the country. In addition to controlling its bay, San Francisco succeeded in getting a state law passed, effective January 1, 1971, which bans all waste disposal in a semicircle in the Pacific Ocean centered on San Francisco and with a radius of 33 miles. This is believed to be a pioneering measure in the protection of coastal waters.

Legislatures pass laws, but judges interpret them. A series of judicial decisions can shape a law more sharply than any legislation, and the way to get judicial decisions is to institute lawsuits. In the last few years environmental suits have proliferated, doubling every six months. Suits have been filed in California to prevent commercial development of a public recreation area, in New York to prevent the building of the Hudson River Express Highway, in Virginia to prevent the "giveaway" of valuable wetlands. Woodsmen are enjoined to spare trees, oil men to protect the seas, chemical plants from polluting the air. A whole new field of law is being developed by ecology-minded lawyers and law professors. In 1971, environmental law courses will be given in about half of the 170 accredited law schools. One fourth of the 70,000 law students are expected to attend. A leading pioneer in environmental law is a 33-year-old professor, Joseph L. Sax, at the University of Michigan. Professor Sax drafted the first state bill, which has been enacted in Michigan, to guarantee the right of individuals to sue any public agency or private industry which is damaging the environment. Similar bills drafted by Professor Sax are now before Congressional committees.

Another type of lawsuit which is increasing is the suit for damages as a result of pollution—including the sound pollution of sonic booms that result in broken windows and cracked ceilings. Many suits have been filed for damages from mercury poisoning, of which the most spectacular is the one for 4 million dollars on behalf of the children of the Huckleby family in Alamagordo, New Mexico. The children ate meat from a hog fed on mercury-treated grain: Amos became blind, Dorothy lost the power of speech, and Ernestine has been in a coma since 1969. A baby, Michael, born after the mother ate the meat, is believed to have contracted congenital mercury poisoning.

Besides laws and courts, an important component of government is the executive branch—the people who act, and who are called administrators when you like what they do and bureaucrats when you don't. A law is as good as the man who administers it, and it is important for citizens to watch appointments to environmental agencies. There was disappointment when Mr. Nixon dropped Walter J. Hickel as Secretary of the Interior. Mr. Hickel had turned out to be a serious and concerned supporter of antipollution measures. For example, for seven unsuccessful years interested Congressmen and environmentalists had been pressing the Department of Interior to do a survey, really an inventory, of the plants which are polluting the nation's waterways and the nature and amount of pollutants. Mr. Hickel ordered the job done—a significant step in controlling water pollution.

On the plus side of the ledger of the Nixon administration was the appointment of John Sheaffer to serve as science adviser to the Corps of Engineers. Sheaffer is the man who dreamed up the Mt. Trashmore project and sparked the Muskegon County project. The Corps of Engineers has the power and authority to control waste in waterways, and they have been very lukewarm about enforcement in the past. The appointment of Mr. Sheaffer may signify a substantial change in attitude.

Probably the greatest power of a government lies in the

power of taxation. No war can be waged without taxes; no
ordinance can be enforced without them. The kind of
taxes levied and what is done with the money are of tremendous influence in shaping a society.

The power of taxation must be mobilized against pollution. Factories, power plants, refineries, etc., must be taxed
according to the amount of pollution they create. This
approach has been used in Europe, particularly in the Ruhr
Valley, which has the greatest industrial complex in the
world for a given area: steel mills, coal mines, chemical and
pulp factories. For several years, all industries and communities have had to join water purification associations
which tax offenders on the basis of the amount of pollutants given off. The money goes for water treatment. The
taxes have provided a strong incentive for water conservation through the recycling and reprocessing of used water,
as already detailed in Chapter 4.

The purpose of the taxation in ecology is not to make
money, but to make pollution unprofitable. Such taxation
should cover municipal and federal installations which are
usually tax-exempt. While at the national level only Congress can levy taxes, the assessment should be recommended
by an independent commission with its own funding. Pollution taxes should not go into the general tax fund, but
should be earmarked for reduction of pollution, for research on one hand and incentive programs on the other.
It is important that the taxes should be federal taxes to
insure uniform standards, although part of the tax receipts
could be returned to local authorities. Segregation of taxation—earmarking specific funds for specific purposes—is
not in general a good policy since it becomes subject to
special pressures, but it has been used very effectively by

the highway lobby in regard to gasoline taxes, and would be a major step forward in protecting the environment.

The special funds here proposed should in no way be a substitute for the use of general taxes for national environmental programs. What government does with its tax receipts reflects the priorities of the society. If 5 billion dollars a year is spent on putting a man on the moon and only a quarter-billion a year is allocated to aid municipal sewage plants, then a man on the moon has a higher priority than clean water. If you disagree with this order of importance, if you wish to establish other priorities, then you must become politically active.

Since ecologists wish to change existing priorities, they must be involved in politics. In fact, so huge are the sums involved, and so drastic the change necessitated in our priorities, it may be that in the next two decades ecology will equal war and racism as a major social issue.

At the present time, all taxes—federal, state and local—take up some 27 per cent of our gross national product. The federal government takes the lion's share, two-to-one, in relation to state and local governments. This was not always the case. In fact, until World War II, the situation was reversed. In 1913, state and local taxes were six times as large as federal taxes, and as late as 1940 they were still twice as large. The war abruptly reversed this position, and from 1940 on, the federal government has taken two thirds of all taxes, primarily because of armaments expenditures.

Even if armaments are reduced, the ratio of two-to-one between federal taxation and local taxation will probably remain for two reasons: first, the federal government can more easily collect money and distribute it in grants to local areas to help the poorer areas, and secondly, the states and localities are reaching a limit to their taxing powers. If New York City raises taxes too high, people will move out. Few people will move out of the country if the federal government raises taxes. Americans think they are heavily taxed, but this is not so. Other countries place a greater burden of taxation on their citizens. A story from Jerusalem (*New York Times*, January 5, 1971) states that Israeli citizens pay 41 per cent of the gross national product in taxation of all kinds, compared to 39 per cent for Sweden, 33 per cent for England, and 27 per cent for the U.S.

The significance of the two-thirds federal take of taxes lies in the fact that ecologists cannot look to state and local governments for massive funds to protect and enhance the environment. Local bodies are too poor—desperate for funds to maintain essential services: fire and police departments, health, education, welfare. There is hardly a city in America that is not in serious financial trouble. Hence, while state and local governments can help with antipollution regulations and some substantial investments, the massive help must be federal. That's where the money is to be found: in the federal budget.

When we look at the latest federal budget (1971–1972) of 229 billion dollars, we find that 40 per cent is legally determined and untouchable; 19 per cent is politically untouchable, being made up of long-term programs in agriculture payments, health and education programs, highway construction, and so on; 37 per cent goes to defense and allied areas, space programs, atomic energy programs, international aid. Only 4 per cent goes to general government departments. There is not much fat in that 4 per cent, and even if we add the politically untouchable programs, a 5 per cent reduction would yield less than 3 billion dollars

WHERE TAX MONEY GOES

1971–72 budget of $229 billion

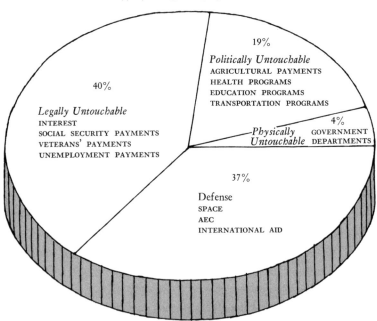

19%

Politically Untouchable
AGRICULTURAL PAYMENTS
HEALTH PROGRAMS
EDUCATION PROGRAMS
TRANSPORTATION PROGRAMS

40%

Legally Untouchable
INTEREST
SOCIAL SECURITY PAYMENTS
VETERANS' PAYMENTS
UNEMPLOYMENT PAYMENTS

4%
Physically GOVERNMENT
Untouchable DEPARTMENTS

37%

Defense
SPACE
AEC
INTERNATIONAL AID

and cut into essential services. The place to look for funds is in the swollen and wasteful arms and space budget, which has ranged between 80 and 90 billion dollars for the past several years. The Vietnam war alone has cost over 100 billion dollars thus far. A realistic approach to pollution would cost some 20 billion a year for the next 10 or 15 years. Therefore, it is clear that the central question of priorities for the American people is whether they want more Navy, more Air Force, more Army, more atomic weapons and more war, or whether they want clean air, clean water, clean food, and clean cities. If they wish to have new priorities, they will have to engage in serious politics and elect officials—presidents, governors, legislators —who will share their concern.

Politics is more than voting at stated intervals. Good men may be elected, good laws passed, fine regulations promulgated, but if they are not enforced, all the work done is useless. Vested interests fight back with counter-suits, evasion, procrastination, and lobbying for modifica-tions of regulations, stretching definitions in their favor, and so on.* Their activities go on across the nation, and as of January 1971, "not one grain of dust, not one liter of gaseous pollution has yet been removed from the atmos-phere of this nation as a direct result of the 1967 air legisla-tion." The quote is from William Ruckelshaus, the Federal Administrator of the Environmental Protection Agency.

An example of delaying tactics, by no means atypical, was given by Congressman Ken Hechler of West Virginia, who is pressing for the adoption of a federal law similar to the Michigan statute allowing citizens to bring en-vironmental lawsuits. It concerns a major corporation, Union Carbide, which has a metallurgical and power plant complex at Marietta, Ohio, close to the West Virginia border. The complex uses soft coal with a high sulphur content, and from its inception in 1952 was a heavy pol-luter of the area. After many unanswered complaints, a group of citizens got Congressman Hechler and Senator Randolph to meet with them on Labor Day, 1965, and as a result the U.S. Public Health Service began an investiga-tion. They estimated that the Marietta plant emitted 44,000 pounds of particles a day, although Union Carbide claimed only 17,000 pounds of emissions. The company, however, refused to allow the investigators to enter the plant or discuss the discrepancy, despite two meetings set up for the purpose. Finally a letter from the Pollution Control Agency was sent to the plant management asking permission to in-spect the plant and requiring certain technical information.

---

* For a thorough discussion of such activities see James Ridgeway, *The Politics of Ecology*, E. P. Dutton & Co., New York, 1970.

The letter was sent on August 27, 1967. Within 10 days management sent a one-paragraph letter saying the agency letter was receiving "careful consideration." Four months passed and nothing happened. On January 11, 1968, the Agency reiterated its requests. On January 31 the plant management replied, evading the request. In April the government requested a meeting with headquarter officials, and the meeting was held on April 25. There Union Carbide people stated that the required information had already been sent to the Ohio State Department of Health and could be obtained there. A letter was sent immediately, answered six weeks later by Ohio, saying they had not received any such information. On July 30 the Pollution Agency threatened legal action unless Union Carbide provided the information at a federal-state antipollution conference to be held in a reasonable time. The company procrastinated; the conference was held a year later, October 30, 1969; Union Carbide didn't show up. Meanwhile a supplemental government report showed that pollution from the Marietta plant had increased since 1967. On December 31, 1969, the Pollution Agency informed Union Carbide of its findings and stated that a failure to report within 30 days would subject the company to fines of $100.00 a day until the information was furnished.

On January 29, 1970—two days before the fines went into effect—Union Carbide sent the information—more than four years after the investigation had begun! Moreover, Union Carbide was supposed to submit an abatement schedule, but on February 17, 1970, the board chairman of Union Carbide wrote the government that it couldn't submit the schedule until its capital budget had been made up. By July 5, 1970, when Congressman Hechler gave the story in detail to the *New York Times,* Union Carbide had still not submitted a pollution-abatement plan. By the end of the month, when it did submit a plan, the federal agency found it was unacceptable with regard to stack heights and control of sulphur oxides. The agency ordered a reduction of 40 per cent in sulphur dioxide by October, 1970, and a 70 per cent reduction by April, 1972. On January 18, 1971, Union Carbide said it would cut emissions by 40 per cent but could not meet the 1972 deadline and would lay off 500 production workers, a course of action which Ralph Nader termed "environmental blackmail" and which Senator Muskie promised to investigate.

This case history is given at length to show how laws and agencies can be thwarted. In part this problem can be met by stronger laws: the new Muskie Bill provides jail sentences as well as stiffer fines for company officials who evade the law. But basically, the vigilance of Congressmen and of the general public is essential, for it is well known that regulatory agencies tend to become allied with the industries they are supposed to regulate.

Since citizens' pressure is intermittent, while industries have well-paid lobbyists, it is necessary for concerned citizens to join those ecologically oriented organizations which keep members informed, educate the public, pinpoint political pressures and have *their* lobbyists who work to protect the environment.* On a local level, just a few citizens can start very effective groups, such as Chicago's SAVE (Society Against Violence to the Environment), or Pittsburgh's GASP (Group Against Smog and Pollution), or the Florida Defenders of the Environment, who suc-

---

* *Ecotactics.* Pocketbooks, New York (95¢), and *The Ecological Handbook*, Ballantine Books, New York (95¢), give information on groups and activities.

ceeded in barring a projected airport in the Everglades and halting permanently the construction of the Cross-Florida Barge Canal. This last victory was a real achievement, since the federal government had already spent 50 million dollars, and one third of the canal had been built. When President Nixon announced the stop action on January 19, 1971, he was undoubtedly mindful that two pro-canal politicians had been retired by the voters in 1970: Governor Kirk (a Republican) and Senator Holland (a Democrat).

Just a couple of people can make a difference. In Washington, Mrs. Richard Helms, wife of the CIA Director, and Mrs. Paul Ignatius, wife of the former Secretary of the Navy, organized a group, Concern, Inc., to focus on consumer shopping from an ecological point of view. The group sends out "Ecotips" which recommends or condemns consumer products. Within a very short time, some of the companies affected began to take notice and to assure the group that they would take remedial action. A particularly effective method of political pressure lies in those areas where the federal government has direct responsibility.

Federal agencies should be setting an example to the nation by devoting time, money, and people to research and application in ecological matters. A notable case of leadership has been provided by the Tennessee Valley Administration which has stressed its role as a proving ground for new antipollution technology. Its program for fiscal 1970 includes the removal of all fly ash from TVA coal-fired plants by 1975, the removal of sulphur dioxide from stack emissions, computerized studies for stopping thermal pollution, and a floodway design along streams to protect fish and wildlife.

An interesting and potentially powerful innovation in

antipollution politics was pioneered in 1969 by a Washington-based nonprofit corporation called Project on Corporate Responsibility. Its first effort, Campaign General Motors, was an attempt to obtain shareholder approval through the solicitation of proxies of several resolutions, including one involving GM's efforts regarding air pollution. (GM spent 250 million dollars on advertising and 15 million on antipollution research in 1969.)

Campaign GM's major success in soliciting institutional proxies came when the New York City Pension Funds voted for its proposals at their annual meeting. Several other municipal and state pension funds also supported one or more of the proposals.

In 1971, there was a second round in Campaign GM. Three proposals to make the company provide detailed information on minority hiring, air pollution, and auto safety policies of GM were passed. A number of prominent institutions voted for these proposals.

Since many municipalities and practically all universities hold blocks of stock in the largest corporations, the technique used by the Project on Corporate Responsibility could be of great value both as a means of educating the public and as a pressure on corporate management.

Successful political action depends on good organization, but basic to organization is education. More and more across the nation, citizens concerned with ecology have been organizing and educating the public. The understanding of ecology has grown tremendously in the last two years, and 1970 may be considered a turning point in the politics of ecology. Early in the year, on April 22, a national Earth Day was so successful that it gave a strong push to politicians. By the end of the year the Muskie Bill was law (to be exact, the President signed the bill into law on January 1, 1971), and this bill is a landmark in ecological politics, for it provides 1.1 billion dollars over three years for implementation and research.

Public consciousness of pollution and its remedies has grown amazingly in the last few years. Increasingly, it is influencing legislators and public officials, as was shown most dramatically in the vote on the SST. As the *New York Times* reported after the vote, "The public cast the deciding vote," and went on to say: "It was widely agreed that the critical, possibly historic point about the SST debate was that a delicate question of public policy and industrial evolution had been taken from the boardrooms and the subcommittee chambers and translated into a broadly political issue . . . and that a new model of decision-making had been fashioned in the process." (*New York Times*, March 28, 1971.)

With antipollution technology on the increase, with education fostering public consciousness and with the political climate favorable, the future for the United States looks more hopeful today than at any time in the past. Americans have it in their power to clean up their country within a decade.

● 15 ●

ECOLOGY AS WORLD POLITICS

Wherein we move from domestic to international problems and solutions. New types of organizations for ecological studies. Warning systems. The problem of poor and rich countries. Where is the money to come from? Some figures on armaments.

Thus far we have been speaking primarily about the United States, both because we live here and because we are the greatest polluters on earth. But the major ecological problems are worldwide. Any long-lasting air and water pollution will ultimately affect the atmosphere and the oceans. The problems of pesticides and fertilizers are inextricably related to agricultural production, which is crucial to population problems. When we start to deal with international problems, we move into the realm of world politics, which are extremely complex and totally different from national politics. A basic consideration, never to be overlooked, is that nations are sovereign bodies that do not easily accept restrictions on their actions.

In considering world politics, two sets of generalizations are useful. One is the division of the world between socialist and capitalist nations, and the other the division of the world between industrialized and underdeveloped nations, or the rich and the poor nations.* These two divisions are not equivalent: there are poor and rich socialist nations, poor and rich capitalist nations. This is fortunate for the

---

* There are minor exceptions. Underdeveloped Saudi Arabia and Libya are very rich because of their oil.

future of the world, because if the rich-poor division reinforced the socialist-capitalist division, the resultant conflict would inevitably tear the world apart. Ecological problems supersede or bypass class, race, and ideology, and
make possible, at least in theory, a working-together for
the benefit of all concerned. It is to the benefit of both
capitalist Japan and socialist China that the narrow seas
between them should not be polluted. The same applies to
Russia, Sweden, Poland, Finland, Denmark, and East and
West Germany on the Baltic. Three of these countries are
communist, four capitalist. Because of their political divisions, there has been a lack of cooperation on Baltic pollution,
but now that the Baltic is seriously threatened they are beginning to exchange information. (*New York Times*, September 30, 1970.)

As we have seen, 80 per cent of all pollution comes from
the industrialized nations, and they have both the money
and the technology for a massive antipollution effort. England, for example, has drastically cut its smoke pollution
from about 2.5 million tons in 1951 to 1 million tons in
1968. All industrialized nations are beginning to move on
pollution as shown in a survey of 15 industrialized nations
by *New York Times* correspondents. The survey, published on June 7, 1971, "found that, quite clearly, 'envi-
ronmentalism' is no passing fad . . . In almost every country
surveyed, new laws have been enacted and environmental
agencies set up. Sizable amounts, rising toward a 1-billion
dollar-a-year level for some countries, are beginning to be
spent by government and industries for pollution control."
The survey goes on to give specifics, country by country.

Japan seem outstanding, with good reason. While the per
capita pollution of Japan is much less than that of the

United States (only one-fourth) it is heavily concentrated. All of Japan is the size of California, and Japanese population density is 25 times that of the United States. Moreover, Japanese industrialization is increasing at a rate nearly three times that of the U.S., so that the country is in very deep trouble ecologically. A dispatch to the *New York Times*, July 29, 1970, gives a few horrendous cases.

In a single village area, Minamata, 46 people have died in 20 years from mercury-poisoned fish, dozens have lost their eyesight or gone mad, and many children have been born mentally retarded or physically defective.

Farms near cadmium plants can no longer sell their milk or their rice because the soil has become so impregnated with poisonous cadmium waste that no vegetation is safe for man or beast.

In the picturesque harbor of Tagonoura, once 28 feet deep, the slime from the waste of several paper-making factories has accumulated to such an extent that the harbor is now only 14 feet deep and can no longer accommodate the ocean-going vessels it once did.

The air pollution in Tokyo is so severe that despite the most lavish care, trees and shrubs in the secluded gardens of the Imperial Palace are dying. Worried court officials try to get the Emperor and Empress out of the city as often as possible.

Faced with a profound crisis, the government has moved with energy and speed. Premier Sato set up a national anti-pollution center to coordinate the struggle, with himself in charge and the Minister of State as the executive. The Ministry of International Trade and Industry is spending 14 million dollars in five years to develop an electric car with a speed of 48 miles an hour and a range of 60 miles before recharging batteries. Tokyo and Osaka are making plans for a 10-year antipollution program for their cities to cost 1.4 billion dollars.

It may turn out that Japan will lead the world in antipollution measures and be both a pacesetter and a teacher. Its people are ingenious, industrious, and socially minded. Government and business are very closely interrelated and devoted to long-range planning to a degree matched only by the socialist nations. However, being completely capitalist, Japan's social and economic measures will be such as to be easily emulated by other capitalist nations.

The rich nations must provide the leadership and ultimately some of the finances for worldwide antipollution measures. Meanwhile there is one area of international cooperation which is not too expensive, not too ideological, and not so restrictive of national sovereignty as to preclude fruitful treaties. I refer to the area of research and information, which at this juncture is vital to further ecological progress. The organization most fitted to take the lead is the United Nations.

In the U.N., mankind has the first glimmerings of a world parliament. While the enforcing powers of the U.N. are very slim indeed, the organization does provide a forum for the mobilization of world opinion. More important, from the standpoint of ecology, it also has many specialized agencies, such as the World Health Organization (WHO), the Food and Agriculture Organization (FAO), and many others which are conducting organized studies and programs all over the world.

In recent years, as the "crisis of survival" for mankind has become ever more apparent, U.N. bodies have begun to move. For example, one of its specialized agencies, the

World Meteorological Organization, with headquarters in Geneva, announced on May 19, 1970, a plan to set up stations in some of the most isolated clean parts of the earth to determine how pollution is affecting the entire world and to seek the causes of such planetary phenomena as the drop in temperature of the air since 1940. From five to ten major stations will be set up to be backed up by 150 special stations in polluted areas, as well as 140,000 weather stations at present reporting to the world organization.

There is no place in the continental United States sufficiently clean to qualify for a major station. This was the considered opinion of Mr. Robert A. McCormick, the American chairman of the advisory committee of the World Meteorological Organization.

A more ambitious plan, for a global network for monitoring all aspects of the environment, moved forward substantially during 1970. The plan was initiated in 1968 by the International Biological Program which set up an *ad hoc* committee of three scientists—an American, a Russian, and a Swede—to study the problem and make proposals. Their interim report was circulated to a joint U.S.-U.S.S.R. task force meeting at the National Academy of Science in Washington, D.C., in February of 1970.

The global monitoring system would comprise 20 major stations in representative localities around the world. Such localities would include a mid-continent grassland, a desert, an alpine environment, coastal sites, and some mid-ocean stations on ships.

These major environmental stations would rely heavily on computers and on data from existing monitoring systems in weather, radiation, solar energy, pollution, and so forth.

They would develop programs and facilities, including earth satellites and the use of "sentinel" plants and animals which are highly sensitive to pollution: for example, radishes are sensitive to chlorine, petunias to sulphur dioxide, some strains of tobacco to ozone, and so forth.

The plan envisages a close watch on the pool of human genetic materials to see how it is affected by the deluge of new chemicals. Every year 300,000 samples of human blood would be studied for evidence of genetic defects, the information to be fed into computers. There would be close study of the effects of pollution on agriculture and on wildlife, both plant and animal.

By far the most important stimulus to research and international cooperation was a conference announced by the United Nations in early 1970. This is nothing less than the First International Conference on the Human Environment to be held in Stockholm in 1972 with some 1,200 scientists and diplomats in attendance. Planning began and in September the Canadian government made available to the U.N. a deputy minister, Mr. Maurice Strong, to be principal environmental adviser to Secretary-General U Thant. Mr. Strong, once head of the Canadian International Development Agency, took up his U.N. post on January 1, 1971 and was assigned to be the chief executive officer of the 1972 conference. By September of 1971 the third of four preparatory conferences had been held, including one in Switzerland which dealt with the conflict between economic development and environmental safety. Also in preparation, the Massachusetts Institute of Technology was host to a meeting of scientists and administrators at Williamstown, Massachusetts, in the summer of 1970. The conference produced many valuable studies, from which this book has benefited. Another preparatory confer-

ence took place on the island of Malta, also in the summer of
1970, sponsored by the Center for Democratic Studies and
focusing its attention on oceanic pollution. Again many valuable studies were prepared for the conference. Unquestionably, many studies are going on all over the world which
have not yet found their way into the world's press.

Partly under the stimulus of these conferences, a new
body was set up in January, 1971: the International Insti-
tute for Environmental Affairs with Robert O. Henderson,
chairman of Atlantic Richfield Company, as chairman, and
Joseph Slater, president of the Salk Institute, as president.
It will conduct an annual 10-week conference at Aspen,
Colorado, and is aiding the U.N. in preparing for the 1972
World Conference.

As conferences on environment have taken place and
various plans put forth, it has become increasingly clear
that a high-powered international agency is essential. In
the March, 1970, issue of *Foreign Affairs*, Mr. George
Kennan, former U.S. Ambassador to the U.S.S.R. and
former chief of the State Department Planning Staff, called
for the establishment of an International Environmental
Agency staffed by scientists and engineers who would be
"true international servants," not bound by the national
interests of individual governments. Almost simultaneously,
on March 14, 1970, an international conference on water
pollution, meeting at Columbia University, New York
City, called for the establishment of such an agency. One
of the conferees, Professor Richard R. Baxter of the
Harvard Law School, said that the proposed International
Environmental Authority would deal with the "world's
stocks of water and air and, in general, with the maintenance of the quality of our world. That mandate may

before long have to be extended to the protection of all the
earth's natural resources, which will increasingly be in
short supply." Eventually, he said, the authority might be
responsible "for the vexing problem of limiting the amount
of human beings on earth."

A word of caution is here necessary. International nego-
tiations are always lengthy, delicate, and complex, particularly in the creation of supra-national agencies. Leaving
aside the question of lack of knowledge on environmental
problems, there are many differences of opinion of great
validity. Progress will take place slowly. Yet there are
encouraging signs. At a closed meeting of the Economic
Commission for Europe, a U.N. body, in Prague during the
first week of May 1971, the two hundred delegates representing both capitalist and socialist countries set up a panel
of senior advisers within the commission to coordinate
work of various environmental bodies, and recommended
the establishment of an all-European information system.
However it was the consensus of the conference that it was
premature to establish an international system of pollution
controls. Some countries felt that standards acceptable to
all nations would have to be set too low, thus providing a
loophole for obstructionists in the countries with stricter
standards.

Such caution does not mean lack of awareness of the
problem. The International Chamber of Commerce, for
example, meeting in Vienna in April 1971, has gone on
record as favoring international standards of pollution. At
about the same time, on the other side of the world,
Chinese Premier Chou En-lai told the Canadian diplomat,
Chester Ronning, that China is conscious of the problem
of water and earth pollution and is studying the experiences

of the more industrialized countries in order to cope with pollution as China industrializes. China's official attitude toward pollution is encouraging. It will take time and effort but in the long run, an international authority on pollution is inevitable. With China's recent admission to the U.N., this is now possible.

Various existing agencies, national and international, could be placed under the proposed authority. The value of such organizations is illustrated by a small but highly effective group at the Smithsonian Institution, known as the Center for Short-Lived Phenomena, which is an international warning system for sudden happenings in the natural world. To qualify, a phenomenon must be natural rather than man-made (although oil spills qualify); must be unpredictable (eclipses don't count); must provide opportunity for scientific research in the field. In addition, the center reports what it calls Urgent Archeological and Urgent Anthropological Events—discoveries which might be obliterated and need immediate study. For example, the Chinandega Footprints, a set of prehistoric footprints found in volcanic ash on a Nicaraguan roadway being built, called for urgent investigation.

Speed is the center's most valuable service, the kind of speed that enables a scientist to travel to the site of a meteor shower while the fragments are still radioactive, or to the site of a volcano while it is still erupting. The center grew out of the communications network for tracking satellites which the Smithsonian already had in operation. Based on this operation, the center began in 1965 with a staff of one man in one room. It now has a staff of eight and an annual budget of $145,000. It has a roster of 2,025 scientists as correspondents and subscribers in 122 countries.

An International Ecological Authority would have two serious political problems to face, over and above the usual national frictions. The first is the conflict between capitalist and socialist nations with the tremendous armaments race it has abetted. The armaments race is not only hugely wasteful of materials and men, but the production of its components—aluminum, steel, chemicals, etc.—adds appreciably to world pollution. The ways and means of disarmament are beyond the scope of this book, but perhaps the "crisis of survival," by requiring scientists of all nations to work for the common good, may soften the ideological conflict and help smooth the way to disarmament.

Even more acute than the problem of disarmament is the division of the world between rich and poor nations, the industrialized and the underdeveloped nations. At the Columbia University Conference mentioned above, a scientist stated the case for the underdeveloped countries: "We've got lots of clean air; we want more smokestacks in our country."

We have seen how in the case of DDT, poor countries refuse to prohibit its use because of their need for food and for disease controls. China refuses to stop testing atom bombs; India wants artificial fertilizers and steel mills. Their moral position is a powerful one. They say to the developed countries: "You have heedlessly polluted the world and *are still doing so* at a greater rate than we are. You've got your atom bombs, your power, your factories, your refineries, your railroads, your automobiles. Now you wish to cry halt! No, thank you; thank you, no! We need to industrialize, and we will industrialize."

If underdeveloped nations are to stop using DDT, the industrialized nations must substitute biological controls—

free. We must develop and supply nonpolluting energy sources—free. We must develop and supply new techniques of industrialization—free.

All this will cost money, a great deal of money, and Americans may say: "Why should we help those countries when we have so much poverty here at home? Let us take care of our own first; let us wipe out poverty here at home." This writer couldn't agree more, but the poverty in the United States is not due to lack of wealth in the country. Paradoxically, money alone will not solve the problem of poverty.

The primary cause of poverty in the United States is lack of minimal education and of skills which command a living wage. This poor education is itself the result of social and economic causes rooted in history: the aftermath of the Civil War in the South, the mass immigrations in the north, the blocking of mass trade unions until the 1930's, the everlasting racial discrimination both north and south, the low wages paid to the parents of the unskilled, and so on. With the proper social policies, the proper social priorities, poverty can be remedied in our nation at comparatively little cost. In fact, since the poor are nonproductive while consuming a lot of social services, the country as a whole would be greatly enriched once the poor fitted into productive life. In comparison, the cost of aid to underdeveloped countries is a small amount.

The compelling reasons for industrial nations helping the underdeveloped countries fall into two main categories, one moral and the other selfish.

The moral reason is that the rich nations owe reparations to the poor nations. They became rich by exploiting them. The main reason poor nations are poor is that most of them

have been colonies under the economic and political control of the imperial powers. For two centuries, their wealth
has been drained to help build the industries of the rich
countries.

Modern industrialization began with the Industrial Revolution in England around the 1770's. The major part of the
capital for the Industrial Revolution came from the British
slave trade and the conquest of India, followed by massive
looting. American industrial development came from capital accumulated mostly in the slave trade and from loans
from England. All industrial nations were helped by the
English Industrial Revolution, either in capital, technology,
or both. After the initial looting of colonies the exploitation
continued, decade after decade, either by direct colonial
rule, as in India, Indochina, Africa, etc., or by the massive
use of economic power, as in Latin America. Because the
wealth of these colonized nations was drained away, they
could never get started on industrialization, but remained
suppliers of raw materials—copper, tin, oil, rubber, etc.—
or agricultural products—coffee, tea, sugar, etc. The profits extracted were enormous: President Roosevelt once got
the British Lord Runciman to admit that England had
taken a 2,000 per cent profit in its colonial rule of African
countries.

The second reason for helping underdeveloped countries
is one of self-interest—what might be called enlightened
selfishness. If there are famines, there will be wars and
revolutions. Any one of these may escalate to involve the
major powers with the potential of thermonuclear war. If
there are famines, there will be epidemics and pestilences.
Any one of these may get out of control and sweep the
world, as influenza did at the end of World War I. If

there are wars and revolutions, the United States will not get its raw materials to keep our economy going. If there are wars and revolutions, we may find an overpopulated Latin America marching north to relieve its pressures. Even if the United States does not get involved in war, it may suffer from the wars of others. France, England, Russia, and China already have nuclear weapons; Israel may already have them or the plants to make them. India may decide to develop them; so may Egypt. The population increase may result in an India-China War, or a Russian-China War, or an Israel-Egypt War—the possibilities are extensive and not at all farfetched. Meanwhile, pollution of the earth is steadily on the increase.

I repeat: industrialized nations can both stop pollution and help industrialize underdeveloped countries at no hardship to themselves. Here is the proof.

Experts have estimated that underdeveloped countries can be industrialized to European levels at a cost of 35 billion dollars a year for 10 years. Of that 35 billion the nonindustrial countries can provide about 10 billion themselves, and the other 25 billion would come from the rich countries. That's a total of 250 billion dollars under existing industrial techniques.

Those techniques create pollution. It has been estimated that it will cost the United States about 20 billion dollars a year for 10 years to remedy the damage done and prevent further damage. So let us assume that pollution-free industrialization would add 100 billion dollars to the cost of industrializing the underdeveloped countries. That makes 350 billion dollars, plus our own 200 billion, plus another 100 billion for preventing pollution in other industrial countries, a total of 650 billion dollars, or 65 billion a year.

The world has already wasted 2,000 billion dollars in armaments in the last 20 years, an average of 100 billion a year. One half of that expenditure was by the United States, one quarter by the U.S.S.R., and one quarter by all other nations, including some of the poorest. In other words, the world has thrown away three times the amount of money sufficient to have the world industrialized and pollution-free on a European standard of living!

Historians a thousand years hence, if they exist, will find it hard to believe that cultured, sophisticated, intelligent human beings could be so blind. They will dig hard into our civilization to find how our institutions have operated to become self-destructive.

The estimates given above are conservative: the world is currently spending on armaments between 150 and 200 billion dollars a year. The United States alone is spending some 80 to 90 billion, of which 30 billion is for Vietnam. That 30 billion, matched by other nations, would be enough for pollution-free industrialization of the world.

It is utopian to think that all armaments could be abolished overnight. It is not utopian to say that they could all be cut by a third within a year and that money devoted to stopping pollution and starvation in the world. This is a feasible political achievement—and what a magnificent subject for politics! Here is a challenge worthy of our best talents, one to which American youth can respond with no regrets and no reservations. We have been told again and again that our country is the most powerful on earth, and that power carries responsibility. So it does. But instead of trying to bomb people into submission, it is time we tried leadership for cooperation. On the Spaceship Earth, only teamwork will bring safety and security.

◖ EPILOGUE ◗

A book is like a child. It is conceived in eagerness and happiness, brought forth in labor and pain, nurtured and matured in love and discipline, sent forth into the world with trepidation, sadness in parting, and a degree of relief. Writing books, like raising children, is, or should be, an educative process. Authors or parents who learn nothing in the process produce neither good books nor good children.

This book and my four children have been closely connected. I wrote it for them, and by extension, for all other children. As I read and studied what past generations have done to the earth, and the resistance to change from the adults in power today, I was often filled with a sense of despair. As I saw my own children reject aggression, self-advancement, and the tyranny of material things, and choose empathy, working together, and the freedom of modest living, I was filled with a sense of hope.

A large and growing section of youth is joining that small but ever-present group of older people which has insisted that the quality of life is more important than the piling up of things; that aggression and cut-throat competition are destructive, and that the earth should be protected. It is just barely possible that this coalition, developing in all countries, may yet save the world.

The spirit of cooperation has always existed in mankind, particularly in the family unit, and becomes overriding in emergencies—floods, earthquakes, epidemics, and so on. Without cooperation, the species could not have survived. In modern times, a misapplication of Darwinism (the survival of the fittest) reinforced an economic and social system that placed a premium on competition. Man is neither wholly competitive nor wholly cooperative. He is a shifting mix of both qualities, and it is society which is decisive in fostering one or the other. Experiments in summer camps have shown conclusively that institutions (camp housing, camp rules, camp rewards, etc.) can develop competition to the point of aggression and bitter divisions, or cooperation to the point of fraternalism and free sharing.

The one thing I've learned in this book is that the human race is doomed if it doesn't transcend its divisions, whether of race, sex, age, ideology, religion, or wealth. There is no way to deal with the enormous problems presented, except by everyone working together. In that great slogan of 1776: "We must hang together or we shall hang separately."

So let this book go out in trepidation, sadness, relief, and some measure of hope.

BIBLIOGRAPHY · INDEX

◖ BIBLIOGRAPHY ◗

Asimov, Isaac. *The World of Carbon.* New York: Collier Books, 1966.

Bernal, J. D. *World Without War.* New York: Monthly Review Press, 1963.

Brittain, Robert. *Let There Be Bread.* New York: Simon & Schuster, 1952. *Rivers, Man and Myths.* New York: Doubleday, 1958.

Brown, Harrison. *The Challenge of Man's Future.* New York: Viking, 1967.

Brown, Lester R. *Man, Land and Food.* FAE Report #11, U.S. Dept. of Agriculture, 1963.

Caldwell, Lynton Keith. *Environment.* New York: Natural History Press, 1970.

Carson, Rachel L. *The Sea Around Us.* New York: Oxford University Press, 1950. *Silent Spring.* Boston: Houghton Mifflin, 1962.

De Bell, Garrett (ed.). *The Environmental Handbook.* New York: Ballantine Books, 1970.

Dassman, Raymond F. *The Destruction of California.* New York: Macmillan, 1965.

Devons, Samuel (ed.). *Biology and the Physical Sciences.* New York: Columbia University Press, 1969.

Ehrlich, Paul R. *The Population Bomb.* New York: Ballantine Books, 1968.

Eisely, Loren. *The Immense Journey*. New York: Random House, 1946.

Elton, Charles S. *The Ecology of Invasions by Animals and Plants*. New York: John Wiley, 1958.

Engel, Fritz-Martin. *Creatures of the Earth's Crust*. London: George C. Harrap, 1965.

Fortune Magazine. *The Environment*. New York: Harper & Row, 1970.

Gamow, George. *The Biography of the Earth*. New York: Viking, 1959. *The Creation of the Universe*. New York: Bantam Books, 1965.

Godfrey, Arthur (ed.). *The Arthur Godfrey Environmental Reader*. New York: Ballantine Books, 1970.

Goldman, Marshall (ed.). *Controlling Pollution: The Economics of a Cleaner America*. Englewood Cliffs, N.J.: Prentice-Hall, 1967.

Graham, Frank Jr. *Disaster by Default*. New York: Modern Library Editions, 1966.

Hecht, Selig. *Explaining the Atom*. New York: Viking Press, 1955.

Helfrich, Harold W. Jr. (ed.). *Agenda for Survival*. New Haven: Yale University Press, 1970. *The Environmental Crisis*. New Haven: Yale University Press, 1970.

Linton, Ron M. *Terracide*. Boston: Little, Brown and Co., 1970.

Maury, Marion. *The Good War*. New York: Macfadden-Bartell, 1965.

Melman, Seymour. *Our Depleted Society*. New York: Holt, Rinehart and Winston, 1965.

Mills, Mark M. and Mainhardt, Robert (eds.). *Modern Nuclear Technology*. New York: McGraw-Hill, 1960.

Mitchell, John G. and Stallings, Constance L. (eds.). *Ecotactics: The Sierra Club Handbook for Environmental Activists*. New York: Pocket Books, 1970.

Nader, Ralph, *Unsafe at Any Speed*. New York: Grossman, 1965.

Oser, Jacob. *Men Must Starve.* New York: Abelard-Schuman, 1957.

Piper, A. M. *Has the United States Enough Water?* U.S. Geological Survey, 1965.

Ridgeway, James. *The Politics of Ecology.* New York: E. P. Dutton, 1970.

Rienow, Robert and Leona. *Moment in the Sun.* New York: Dial Press, 1967.

Shepard, Paul. *Man in the Landscape.* New York: Alfred A. Knopf, 1967.

Shepard, Paul and McKinley, Daniel (eds.). *The Subversive Science.* Boston: Houghton Mifflin, 1969.

Shurcliff, William A. sst *and the Sonic Boom Handbook.* New York: Ballantine Books, 1970.

Smith, Homer W. *From Fish to Philosopher.* Boston: Little, Brown and Co., 1953.

Storer, John H. *The Web of Life.* New York: Devin-Adair, 1956.

Whyte, A. Gowans. *The Ladder of Life.* London: C. A. Watts, 1951.

Wolfert, Ira. *An Epidemic of Genius.* New York: Simon and Schuster, 1960.

Williamstown Conference, *Man's Impact on the Global Environment.* Cambridge, Mass.: MIT Press, 1970.

United Nations. *Possibilities of Increasing World Production.* fao Basic Study #10, New York: United Nations, 1963.

Periodicals: The periodicals most frequently used were *Environment, The Scientific American, The Bulletin of Atomic Scientists,* and the *Saturday Review,* which are all listed in the Reader's Guide to Periodical Literature. Other periodicals used are given in the text.

❶ INDEX ❶

226

⏺ ABOUT THE AUTHOR ⏺

Carl Marzani was graduated from Williams College, Magna Cum Laude. After receiving the Moody Fellowship to Oxford University, he was graduated with honors in the school of Modern Greats (Philosophy, Politics, and Economics).

He has taught economics at New York University and has lectured at many universities throughout the years. During World War II, he served in the Office of Strategic Services and after the war, held a responsible position in the State Department Intelligence Office until he resigned in protest against the developing Cold War.

Since then he has had several careers as trade union official, film producer, publisher, and real estate developer. He has written a number of books on economics and politics. Thus he brings to ecological studies the practical sense of a man of affairs as well as the academic competence of several disciplines.

Mr. Marzani is married and lives in New York City. He recently became a grandfather, an event which made his youngest son an uncle at the age of two months.